How and Why

IN BASIC MECHANICS

Arvind Kumar
Shrish Barve

Universities Press

To

The Unknown Physics Teacher

Universities Press (India) Private Limited

Registered Office
3-6-747/1/A & 3-6-754/1, Himayatnagar,
Hyderabad 500 029 (A.P.), India
e-mail : info@universitiespress.com

Distributed by
Orient Blackswan Private Limited

Registered Office
3-6-752, Himayatnagar, Hyderabad 500 029 (A.P), India

Other Offices
Bangalore / Bhopal / Bhubaneswar / Chennai Ernakulam
Guwahati / Hyderabad / Jaipur / Kolkata /Lucknow / Mumbai
New Delhi / Patna

First published 2002
Reprinted 2013

ISBN: 978 81 7371 420 7

Typeset by
OSDATA, Hyderabad 500 029

Printed in India at
BlueBluebells Secure Prints
Hyderabad 500 072

Published by
Universities Press (India) Private Limited
3-6-747/1/A & 3-6-754/1, Himayatnagar,
Hyderabad 500 029 (A.P.), India

Contents

	Preface	v
1.	Units and Dimensions	1
2.	Scalars and Vectors	22
3.	Kinematics	53
	I. One-dimensional Motion	53
	II. Two- and Three-dimensional Motion	79
4.	Laws of Motion	90
5.	Newtonian Relativity	114
6.	Energy and Momentum	150
7.	Systems of Particles	185
8.	Rigid Bodies	215
9.	Illustrative Problems	241
	Index	288

Preface

The imagined student in this book is concurrently going through a standard introductory course in mechanics. The student is familiar with most parts of the course and has also practiced solving problems. Yet she finds many points unclear and elusive, and she is keen on understanding them critically. The teacher respects the spirit of the student, argues with her on equal terms, and attempts to clarify the points and offer additional insights. The result is a dialogue through different topics of mechanics, sometimes dwelling on a minor confusion, other times discussing a major conceptual hurdle. The teacher tries to explain with restraint, and judiciously leaves a point when further going might be too technical or advanced.

We believe readers who share the student's doubts on this or that point in introductory mechanics will find the dialogue useful, and gain some fresh clarifications and insights. But more important, after the book is over, they, like the insatiable student here, should be ready to ask more questions, and yet more questions.

The work on this book was initiated mainly due to the encouragement of Professors B M Udgaonkar and V G Kulkarni. Prof K Chandrasekharan (formerly of University of Madurai) offered useful comments on the manuscript. Some of the learning barriers in Newtonian Relativity discussed in this book have been investigated in detail in collaboration with our colleagues J Ramadas and S Panse. We record our gratitude to all of them. It is also a pleasure to thank Shri Anand Ghaisas for help concerning the diagrams in the initial version of this book.

1

Units and Dimensions

Base Dimensions

S1 *I am sometimes puzzled by the phrase 'dimensions of a physical quantity'. Does the word 'dimensions' refer here to some kind of size?*

T1 No. In ordinary language the word 'dimension' usually refers to size, and this is the sense in which it is often used in physics also, as in 'radiation of wavelength large compared to atomic dimensions'. However, in a phrase such as 'dimensions of angular momentum' the word has a very different meaning.

Dimensions of a physical quantity are related to the base units required to specify it; e.g. the unit of angular momentum is kg m^2s^{-1}, and the dimensions of angular momentum are written as $[M]\ [L]^2\ [T]^{-1}$. In the same way, the unit of resistance in SI is

$$1\ \Omega = 1\ \text{W A}^{-2} = 1\ \text{kg m}^2\ \text{s}^{-3}\ \text{A}^{-2},$$

i.e. the dimensions of resistance are $[M]\ [L]^2[T]^{-3}[A]^{-2}$.

S2 *We also talk of 'one-dimensional motion'; 'three-dimensional space' etc. What does the word 'dimension' mean here?*

T2 The word 'dimension' in these usages means something entirely different from what I said in T1. Here it refers to the number of co-ordinates required to specify the position of a point, that is the number of variables required to span the given space (set of points). Thus motion along a line is one-dimensional because only one co-ordinate is enough to specify the location of a particle at an instant. Space is three-dimensional since we need three co-ordinates to specify the position of any point.

S3 *I do not like this. The same word 'dimension' means different things in different contexts. We know this happens in ordinary language. But in science we should use a separate word for each distinct meaning, to avoid confusion.*

T3 You are right. We indeed exercise this care in science mostly. Different technical terms in science have clear and distinct meanings. Occasionally, however, two or more distinct technical notions happen to be named by the same word. In such cases, the meaning should be clear from the context.

S4 *From T1, dimensions of a physical quantity and its units seem to be much the same thing. Is there any difference?*

T4 The particular units of a quantity depend on the choice of standards for the base quantities of the system. For example, the units of energy are kg m^2 s^{-2} in SI, and g cm^2 s^{-2} in the CGS system. The dimensions of energy in each case are $[M][L]^2[T]^{-2}$. Thus 'dimensions' of a quantity specify the combination of the base quantities (length, mass, time, etc.) required to describe it without specifying the standards for the base quantities employed in the system. In this sense, the 'dimensions' of a physical quantity is a more general notion than the 'units' employed to measure the quantity.

S5 *Does that mean dimensions of a quantity are the same no matter what system of units is employed?*

T5 Not always. If two systems of units employ the same base quantities, but differ only in the standards used for those base quantities, then what you said is true. Thus in mechanics both SI and CGS system have mass, length and time as base quantities. But the units of mass and length (i.e. the standards employed for their measurement) differ: kg and m in SI, g and cm in CGS system. The dimensions of a given mechanical quantity are, therefore, common to both SI and CGS. The units of that quantity may, however, differ. For example, the dimensions of force are $[M][L][T]^{-2}$ in both SI and CGS systems; but the units of force are 1 N = 1 kg m s^{-2} in SI and 1 dyne = 1 g cm s^{-2} in CGS system.

But it is wrong to think that systems of units differ only in the choice of standards for base quantities. Two systems can differ in the choice of base quantities themselves. This is so even for SI and CGS when we go beyond mechanics and consider electromagnetic phenomena. For example, in SI current (or equivalently charge = current × time) is considered a base quantity while in the Gaussian CGS system it is a derived quantity, not a base quantity. Consequently, the dimensions of capacitance, for instance, are $[Q]^2$ $[M]^{-1}[L]^{-2}[T]^2$ in SI and [L] in the Gaussian CGS system. Here you see, therefore, an example of a quantity whose dimensions (and, of course, units) differ from one system to another.

S6 *Can we have a quantity which has units but no dimensions?*

T6 It is possible in a somewhat artificial way. For example, if a clock loses 1 minute each day, we may characterize its rate of losing time as 1 minute/day. This quantity has units but no dimensions. But if we choose a common unit of time for the numerator and the denominator, we can write the same quantity as 1/1440 which has neither any units nor dimensions. As another example, the efficiency of an electric bulb may be expressed as so many lumens/watt; but it can also be expressed by a number (with no units or dimensions) which gives the fraction of input power that appears as luminous radiation.

A more familiar example is angle. Angle is defined to be arc length divided by radius; it is, clearly, dimensionless. Still, it is customary to give it a unit called radian which is defined to be the angle for which the arc length equals radius.

S7 *In SI the five base units m, kg, s, K, and A, do appear to be independent; but the other two units, namely mol and cd, seem to be derived units. Mole is a unit for the amount of substance, but we already have a unit of mass (kg); a candela is a unit for luminous intensity, i.e. luminous flux per unit solid angle. But luminous flux is analogous to radiant power, and we already have a derived unit of power (1 W =1 kg m^2 s^{-3}). What is the point in defining these redundant units?*

T7 A mole does not represent a fixed amount of mass. A mole of C^{12} means a mass of 12 g of C^{12}, whereas a mole of O_2 means a mass of about 32 g. What is common is the number of elementary entities; atoms in case of C^{12} and molecules in case of O_2. In many situations, the number of atoms or molecules in a substance is more significant than the mass of the substance. For example, if the gas law is expressed in terms of the number of molecules; $PV = kNT$, k is a universal constant. If the same law is expressed in terms of mass as $PV = \alpha mT$, α is not a universal constant, but depends on the molecular mass of the gas. Thus it is very convenient to define a unit (mole) which corresponds to a fixed number (Avogadro's number) of elementary entities.

A lumen is a unit of luminous flux. It is defined to be that quantity of light which equals the quantity of light of frequency 5.40×10^{14} Hz having a radiant power of 1/683 W. Many light sources do not radiate isotropically. To characterize the amount of light emitted in a given direction one defines another quantity called luminous intensity which is luminous flux in that direction per unit solid angle.

The unit of luminous intensity is one candela: 1 cd = 1 lm sr^{-1}.†
The reason why we need these different units is two-fold. First, usually only a fraction of the radiant power of a source lies in the visible spectrum, and this fraction may differ from source to source. Second, brightness of an object depends not only on the 'visible' part of the radiant power but is also determined by the sensitivity of the eye to different portions of the visible spectrum. This means two sources having the same radiant power need not appear equally bright to our eyes. Consequently, to measure luminous flux (or luminous intensity in a particular direction) a separate unit is necessary.

S8 *In SI seven base units are defined : metre, kilogram, second, kelvin, ampere, candela and mole. Is there some significance to the number seven here? Does it mean that in nature there are seven fundamental quantities and the rest are all derived from them?*

T8 No doubt you have got this impression from the word 'base quantities'. But your idea is wrong.

The number of base quantities, which particular quantities to take as base quantities, and the standards for the base quantities are all a matter of choice. As already remarked, current is a base quantity in SI but not in the Gaussian CGS Thus if we restrict to mechanics and electricity there are four base units (m, kg, s, A) in SI, but only three base units (cm, g, s) in the Gaussian CGS system.

The flexibility in the number of base dimensions is nicely illustrated when you consider the system of units called "natural units", which are employed in relativistic quantum physics. Here we take $\hbar = c = 1$ (dimensionless) and have only one base dimension in terms of which all the other physical dimensions are expressible.

S9 *I can understand that certain constants like $\hbar$ and c can be assigned value 1 in some system of units, but how can you 'take' them to be dimensionless? $\hbar$ and c have definite dimensions; $[M][L]^2[T]^{-1}$ and $[L][T]^{-1}$ respectively. How do you do away with these dimensions?*

T9 Of course you cannot do away with the dimensions of $\hbar$ and c if you continue to use mass, length and time as base dimensions. The idea is to relate the dimensions of mass, length and time to one another so that $\hbar$ and c become dimensionless.

† Note: Luminous intensity is related to radiant power per unit solid angle. This is unlike the usual definition of intensity which is power per unit area.

S10 *I do not understand that.*

T10 From the relation, speed = length/time, speed is dimensionless, if the dimension of length is the same as that of time. Next, from $E = mc^2$, energy and mass have the same dimension, since c is dimensionless. Further from $E = \hbar\omega = h\nu$ energy, and hence mass, have the dimensions of inverse time (or inverse length). Thus we may consider only one base dimension say length [L]. Dimensions of all quantities can then be expressed in terms of this base dimension. In natural units:
Time : [L], Mass : $[L]^{-1}$, Energy : $[L]^{-1}$, Momentum : $[L]^{-1}$.

S11 *Do I take then that at least one physical base dimension is necessary, no matter what the system of units?*

T11 Not at all. In relativistic quantum physics involving gravity we sometimes take $\hbar = c = G = 1$ (dimensionless), which implies that all the quantities in this system are dimensionless. †

S12 *All this sounds bewildering. What happens to the requirement that physical equations should be dimensionally consistent; i.e. the powers of* [L], [T], [M], *etc. should separately match on both sides of an equation?*

T12 In a system that has [L], [T], [M] as base dimensions, the powers of these three dimensions should match. In a system that has [L], [T], [M], [Q] as base dimensions, the powers of all the four dimensions must match. In natural units if we choose [L] to be the base dimension, the powers of [L] must match on either side of an equation.

S13 *That still sounds very confusing. According to what you have said, an equation like $E = (1/2)mv^3$ is dimensionally wrong in SI or CGS but dimensionally correct in natural units. Isn't there a contradiction?*

T13 There is no contradiction. Remember, a dimensionally wrong equation is necessarily wrong. But a dimensionally correct equation may be right or wrong, actually. In other words, dimensional analysis can only rule out an equation but it can never establish any equation as actually correct. This power to rule out is naturally reduced

† The dimensions of G in terms of [M], [L] and [T] are $[M]^{-1}[L]^3\,[T]^{-2}$ and in natural units they are $[L]^2$. Thus if G is dimensionless so is length and hence, all other quantities.

if the number of base dimensions is reduced. Thus the equation $E = (1/2)mv^3$ can be ruled out on dimensional grounds in SI or CGS; but in natural units you will need to invoke physics to rule it out.

S14 *Let me turn to another point that often bothers me. Sometimes the constant of proportionality appearing in a physical law is simply taken to be unity. For example, Newton's second law is stated as $F \propto ma$, i.e. $F = kma$ where k is a constant of proportionality, and k is then taken to be 1. What is the meaning of taking an unknown thing like k to be 1?*

T14 It is this way. Look at the right side of the equation. The SI unit of mass is kg and that of acceleration is m s^{-2}. The unit of force on the left side is yet to be defined. If I take $k = 1$, the unit of force (newton) is one that produces an acceleration of 1 m s^{-2} in a mass of 1 kg. But I can as well take some different value of k, say $k = 5$. That will simply amount to having a unit of force which is different from 1 newton. Let us call this unit of force as 1 newt. Now $k = 5$ implies that the force required to produce an acceleration of 1 m s^{-2} in a mass of 1 kg is 5 newt, i.e. 1 newton = 5 newt. In short, the value of k simply determines the size of the unit. Therefore, without any loss of generality, we can take $k = 1$ for convenience.

S15 *I encounter this phrase 'without any loss of generality' so often in physics. Please explain what this is supposed to mean.*

T15 Before that you must understand the words 'general' and 'special' as they are used in physics. If a statement or a law is true in all situations, we say it is true in general. If something is true in restricted or particular situations, we say that it is true in those 'special cases'. As an example, the general theory of relativity is applicable to all observers, while the special theory of relativity is applicable only to inertial observers. Another example that may be more familiar to you concerns collisions of two bodies. Here conservation of total momentum holds good in general, whereas total kinetic energy is conserved in the special case of elastic collisions.

Now, while handling a physical situation we may make some convenient choice without restricting or narrowing the problem to some particular situations. We then say that there is 'no loss of generality' in making such a choice. For example, consider the problem of motion of a charged particle in uniform electric and magnetic fields $\boldsymbol{E}$ and $\boldsymbol{B}$. There is no loss of generality in aligning one of the co-ordinate axes, say the z-axis, along either $\boldsymbol{E}$ or $\boldsymbol{B}$. There is also no loss of generality in taking the plane of $\boldsymbol{E}$ and

B to be one of the x-y, y-z or z-x planes. But if we take both **E** and **B** to be aligned along the axes, say **E** along the z-axis and **B** along the y-axis, we would lose generality because we would be restricting the problem to the situations when **E** and **B** are mutually perpendicular.

S16 *Well, let us return to S14. It is clear to me now that we lose no generality in taking $k = 1$. But what about the dimensions of k? When you write 1 N = 1 kg m s^{-2} you are obviously taking k to be dimensionless. Is a constant of proportionality always supposed to be dimensionless?*

T16 Not necessarily. Here we have <u>chosen</u> k to be dimensionless and defined the unit of force via the relation $F = k\ ma$. But we could have defined the unit of force in some other way. For example, we could have defined the unit of force via the Hooke's law $F = \kappa x$ (for some standard spring). Suppose we call this unit 1 hooke (in deference to Hooke who was Newton's rival!). The constant k in Newton's law $F = k\ ma$ will then have the units hooke kg^{-1} m^{-1} s^2.

As another example of a constant of proportionality being dimensional, remember Coulomb's law:

$$F = k\,\frac{q_1 q_2}{r^2}$$

In the Gaussian CGS system, k is taken to be dimensionless, and thereby the dimensions of charge can be expressed in terms of [M], [L] and [T]; $[Q] = [M]^{1/2}\ [L]^{3/2}\ [T]^{-1}$. But in SI, current (and hence charge) is defined independently via Ampere's law. Consequently in SI, k is not dimensionless, but has the dimensions $[M][L]^3\ [T]^{-2}[Q]^{-2}$.

Significant Figures

S17 *Let me move to another topic. The notion of significant figures is often discussed in connection with measurement. What is the meaning of significant figures in a measurement?*

T17 This is best illustrated by an example. Suppose that the result of a length measurement is given to be 13.59 m. We say all the four figures here are significant if the length has been measured to an accuracy of 1 cm i.e. the actual length is established to lie between 13.585 m and 13.595 m.

S18 *But I can write the same number as 13.590 m. That makes the number of significant figures 5. Am I right?*

T18 No. The zero in 13.590 is not significant.

S19 *Why not?*

T19 Because, writing a zero in front of the number will not change the numerical value of the result but it changes the claim as to how accurate the measurement was. When you write the length to be 13.590 you are implying that the length has been measured to an accuracy of 1 mm, i.e. the actual length lies between 13.5895 m and 13.5905 m. This is not the accuracy of the original measurement.

S20 *That confuses me a little. I can always put any number of zeroes at the end of the decimal part of a number. How am I to know how many of these are significant or not?*

T20 Significant figures refer to the accuracy of measurement. When an experimentalist gives a result of measurement, he takes into account the accuracy of his procedure and rounds off the number he gives such that all the figures in the number are significant. Thus if he gives the length as 13.59 m, his measurement according to him is correct up to 4 significant figures, and I have no right to write it as 13.590 m or 13.5900 m etc. because that gives a misleading impression about the accuracy of the experiment. If, however, he quotes the result as 13.590 m, then it is understood that he means the last zero to be significant because he has measured the length to an accuracy of 1 mm. The number of significant figures is then five.

S21 *Does the number of significant figures depend on the units employed?*

T21 No. For example, 13.59 m = 0.01359 km = 1359 cm = 13590 mm. The number of significant figures is four in each case. The number of places of decimal differs. This is why it makes more sense to talk of accuracy of measurement in terms of the number of significant figures than saying that a certain thing is measured correct up to so many places of decimal.

S22 *Can you summarize for me the rules regarding counting of significant figures in a number?*

T22 It is better to go by the simple meaning of significant figures as explained in T20 than to use any rules. Still, if you insist, you can see these rules from the following examples:

The case when there are no zeroes in the number is no problem. A length reported as 27.572 m has 5 significant figures. A time interval given as 5.74 s has 3 significant figures.

A mass reported as 200.007 kg has 6, not 2, significant figures. The zeroes here are significant. One way of seeing this is we cannot get rid of them by multiplying the number by powers of ten.

200.007 kg = 200007 g; both have 6 significant figures.

But a mass written as 0.0007 kg has 1, not 4, as significant figure. Here by multiplying by powers of ten we can get rid of the zeroes. 0.0007 kg = 0.7 g both have only 1 significant figure.

A measurement 1.51 m has 3 significant figures. If it is written as 1510 mm the last zero is not significant. However, if a measurement is reported as 1510 mm or 1.510 m or 151.0 cm, the last zero is significant.

To avoid confusion regarding the extra zeroes which appear when the same measurement is expressed in different units, it is conventional to write the measured number as a number with only one digit to the left of decimal point, multiplied by an appropriate power of ten. For counting the significant figures, then, the powers of ten are ignored. The remaining part is supposed to contain figures which are all significant

Thus 200.007 kg = 2.00007×10^2 kg = 2.00007×10^5 g, all have 6 significant figures.

Similarly, 1.51 m = 1.51×10^{-3} km = 1.51×10^2 cm have 3 significant figures. If, however, a length is specified as 1.510×10^{-6} m then it is meant to have 4 significant figures and not 3.

S23 *What happens if a calculation uses quantities with different numbers of significant figures.*

T23 The answer should be rounded off to the lowest number of significant figures. For example, 5 kg + 2.1628 kg = 7.1628 kg ≈ 7 kg.

4.2 N × 1.532 m ≈ 6.4 J

1525 m/94.3 s ≈ 16.2 m/s

But remember, constants such as 2, π etc. appearing in a formula have no limit on their accuracy. Thus the constant 4 in the formula $A = 4\pi R^2$ should not be thought of as having only 1 significant figure; strictly it has an infinite number of significant figures (i.e. 4 is to be thought of as 4.00 ...). The rounding off of the final result, in this case, will be determined by the number of significant figures in the measured value of R. Also, the value of π should be written correct up to as many significant figures as there are in R. For example, if $R = 1.55$ m then we should take $\pi = 3.14$ which gives $A = 30.2$ m^2. If $R = 1.5539$ m then we should take $\pi = 3.1416$ which gives $A = 30.343$ m^2.

Approximations

S24 *Let us move to another thing that bothers me. We often make approximations in physics and say that something is negligible because it is small. But the smallness and bigness of a quantity depends on the units employed. Then how do we justify such approximations?*

T24 This matter is quite simple. 'Smallness' or 'bigness' is always in comparison to something else, even though this 'something else' is often not explicitly mentioned. In science when you talk of 'smallness' or 'bigness', be sure to consider a ratio of two similar quantities, i.e. quantities having the same dimensions. This ratio is dimensionless and its value will not change from one choice of units to another. 'Smallness' corresponds to this dimensionless ratio being much smaller than 1; 'bigness' corresponds to this ratio being much greater than 1.

S25 *Can you illustrate this point?*

T25 In mechanics when we represent an extended body by a pointlike particle, we consider the dimensionless ratio of the size of the body, s, to the typical distance involved in the motion, d. If $s/d \ll 1$ then the particle approximation is good.† For example, a 100 m long train is approximately a point particle if we are dealing with its motion over hundreds of kilometers since the ratio s/d in this case is $\sim 10^{-3} \ll 1$. In the same way, the dimensionless ratio of interest for a simple pendulum is amplitude divided by the length of the

† Often one is interested only in the translational motion of an extended object. It is then possible to represent the object by its centre of mass, no matter what its size is compared to the displacement.

pendulum. (This is the same thing as the angular amplitude.) The familiar formula $T = 2\pi\sqrt{\ell/g}$ is good in the approximation of this dimensionless ratio being small. Similarly Newtonian mechanics is valid in the approximation of the dimensionless ratio $v/c \ll 1$, where v is the speed of the body under consideration and c is the speed of light. We should try to cultivate the habit of looking for dimensionless ratios in a problem, especially when some approximations are being made.

S26 *You used words like 'this approximation is good under such and such conditions'. But 'good' and 'bad' are qualitative words. Do we not have a precise quantitative way of talking about approximations?*

T26 Surely, there is. A quantitative statement regarding an approximation involves some dimensionless ratio. Let us call this dimensionless ratio α in some particular problem. Suppose some quantity in the problem, say A, depends on α. Mathematically it is usually possible to express $A(\alpha)$ as a power series in α, i.e.,

$$A(\alpha) = A_0 + \alpha A_1 + \alpha^2 A_2 + \alpha^3 A_3 + \ldots$$

For $\alpha < 1$ the successive terms in this series are expected to be smaller and smaller.† This brings out the important point that approximation is a matter of degree (or order as it is called in physics). In the crudest approximation, we can simply ignore all the terms except the first and write

$$A(\alpha) \approx A_0. \qquad \text{(Zeroth order approximation)}$$

By retaining more and more terms on the right side we can go on improving our approximation;

$$A(\alpha) \approx A_0 + \alpha A_1. \qquad \text{(First order approximation)}$$

$$A(\alpha) \approx A_0 + \alpha A_1 + \alpha^2 A_2. \qquad \text{(Second order approximation)}$$

$$A(\alpha) \approx A_0 + \alpha A_1 + \alpha^2 A_2 + \alpha^3 A_3. \qquad \text{(Third order approximation)}$$

And so on. Thus, if in a calculation involving α which is small, we consistently neglect powers of α higher than α^n, the results of the calculation can be claimed to be true up to n^{th} order of approximation in α.

† Strictly, there could be problems of convergence of the series which we will ignore here.

S27 *This seems too technical. Can you explain it with some examples?*

T27 The discussion in T26 may seem a little technical, but I advise you to go over it carefully. The notion of order of approximation is very important in physics. If you are not clear about it, you will tend to look at the approximation procedures in physics as some kind of clever but *ad hoc* tricks to simplify calculations and obtain 'desired' results. That is not so. Approximation is a well-defined quantitative procedure in science. As a first example consider the derivation of the result $f = R/2$ for a concave spherical mirror.

Box 1.1

Consider a spherical mirror with OM as its principal axis. (See Fig. 1.1).

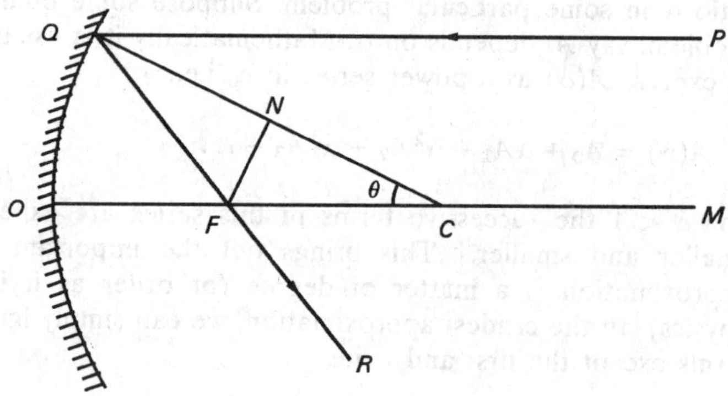

Fig. 1.1 Locating the focus of a concave spherical mirror.

PQ is a ray incident along a direction parallel to the principal axis. QC is a radius of curvature of the mirror. The ray PQ is reflected along QR in accordance with the laws of reflection; $\angle PQC = \angle CQR$. QR intersects the principal axis at F. Since the incident ray PQ is parallel to the principal axis $\angle PQC = \angle QCF$. Thus ΔQCF is an isosceles triangle. FN is drawn perpendicular to QC; clearly $CN = R/2$. Let us denote distance FC by x, consequently $\cos\theta = R/2x$. The distance f between the pole O and F is given by

$$f = R - x = R\left(1 - \frac{1}{2\cos\theta}\right) \qquad (1)$$

This equation may be rewritten using the series expansion of the cosine function as

$$f = R\left[1 - \frac{1}{2}\left(1 - \frac{\theta^2}{2!} + \ldots\right)^{-1}\right]$$

$$= R\left[1 - \frac{1}{2}\left(1 + \frac{\theta^2}{2!} + \ldots\right)\right].$$

While writing the second step we have employed the binomial expansion.

The angle θ depends on how far the incident ray is from the principal axis; its value is maximum for rays incident near the rim of the mirror. If the aperture of the mirror is small compared to its radius of curvature, θ will be small for all the rays. We may then retain only the first term in the expansion of the cosine function to obtain an expression for f as

$$f^{(0)} = \frac{1}{2} R.$$

It is now apparent that this familiar result is only a zeroth order (in θ) approximation to the more general result given in Eq. (1). The superscript (0) on f indicates the order of approximation. A more accurate result, correct to second order in θ, is obtained if we retain one more term in the expansion,

$$f = \left(\frac{1}{2} - \frac{\theta^2}{4}\right) R.$$

Note that only in the zeroth order is f independent of θ. Thus only a beam of ***paraxial*** rays converges to a single point. The notions of focus and focal length, therefore, apply only to paraxial rays.

As another example, let us obtain an approximate expression for kinetic energy from the exact relativistic expression.

Box 1.2

The relativistic expression for the kinetic energy of a particle with mass m is

$$K = \frac{mc^2}{\sqrt{1-\left(\frac{v}{c}\right)^2}} - mc^2.$$

The first term is often referred to as the total energy to remind us that it includes kinetic energy and the so-called rest energy. When the rest energy is subtracted from the total energy we obtain the kinetic energy. The square root in the denominator of the first term can be expanded using the binomial theorem to obtain

$$K = mc^2\left[1+\frac{1}{2}\left(\frac{v}{c}\right)^2+\frac{3}{8}\left(\frac{v}{c}\right)^4+\ldots\right] - mc^2.$$

As is evident, the expansion is in terms of the dimensionless ratio v/c. Now order by order approximations can be made; the lowest order approximation gives us the familiar non-relativistic expression for kinetic energy

$$K = \frac{1}{2}mv^2.$$

Approximate expressions for kinetic energy with more and more accuracy can be obtained by retaining terms in the above expansion up to higher and higher powers of v/c.

In these examples we have started with the exact expression and reduced it to different orders of approximation. Sometimes it is not possible to obtain the exact result at all, but an approximate result may be obtained by neglecting terms involving the dimensionless ratio α relevant to the problem. In such a case the final result can be claimed to be correct up to the nth order in α provided the terms we have neglected contain powers of α greater than or equal to $(n+1)$.

Dimensional Analysis

S28 *Another point. I have read this somewhere that sine and cosine functions in any physical expression must have dimensionless arguments. Is that because sine and cosine are functions of the angle-and this angle is always dimensionless?*

T28 That is true, but I can make a more general statement. Any trigonometric, logarithmic, exponential, Gaussian, hyperbolic functions appearing in physics must have dimensionless arguments.

S29 *How do you see that?*

T29 It will help if you know one piece of mathematics: the series expansions of these functions

$$\sin x = x - \frac{x^3}{3!} + \frac{x^5}{5!} - \ldots$$

$$\cos x = 1 - \frac{x^2}{2!} + \frac{x^4}{4!} - \ldots$$

$$\log(1 + x) = x - \frac{x^2}{2} + \frac{x^3}{3} - \ldots$$

$$e^x = 1 + x + \frac{x^2}{2!} + \frac{x^3}{3!} - \ldots$$

Sinh x and $\cosh x$ being built out of e^x and e^{-x} also have their own series. Do not worry if you are not very familiar with these series. All you need to know is that they involve sums of different powers of the argument x. Now if x has some dimensions, adding to it x^2 or x^3 etc. will not make sense dimensionally. (You cannot add m to m^2 or m^3.) Thus in order for these expansions to make sense, the only possible thing is that x be dimensionless.

S30 *So if an expression in whatever context in physics has a sine, cosine, log, exponential, hyperbolic or Gaussian etc., I can be sure that the argument of these functions has got to be a dimensionless quantity. This is a very interesting point, but is it of any use?*

T30 This is no additional constraint to be imposed from outside. If you do your derivations properly, you will automatically get dimensionless arguments in these functions. You should use this rule as a check on your expressions. Thus if you ever get an expression containing $\sin(t/\omega)$ where ω is angular frequency, you can be sure it is wrong because t/ω is not dimensionless. You must always get the dimensionless product ωt (or its multiple) as the argument in these functions. In the same way if in a correct physical expression you get something like $\exp\{-x^2/a^2\}$ you can be sure that a has the same dimensions as x, because only then is the argument x^2/a^2 dimensionless.

S31 *That reminds me of another little doubt. When I look at an expression or equation in physics, should I take all the numbers such as 2, 1/3, π, etc. as dimensionless?*

T31 It depends. If the expression you are looking at is general in the sense no numbers have yet been put in for its parameters and variables, then any numerical figure that you see is dimensionless. For example, in the expression for kinetic energy, $K = (1/2)mv^2$, the factor 1/2 is dimensionless. You may, however, put $m = 10$ kg in this expression for calculating the kinetic energy of a particular object. The expression then becomes $K = 5v^2$. The factor of 5 in this expression is <u>not</u> dimensionless.

As another example, consider the expression for the displacement of a particle in simple harmonic motion:

$$y = A\cos\left(\frac{2\pi t}{T} + \phi\right).$$

2π in this expression is dimensionless. Now if you put numerical values for A, ω and ϕ in some particular problem, say $A = 2$ cm, $T = 10s$, $\phi = \pi/4$, we get

$$y = 2\cos\left(\frac{\pi t}{5} + \frac{\pi}{4}\right).$$

In this equation you must note that 2 has the same dimension as that of y, i.e. [L]; the number $\pi/5$ has the dimension of $[\mathrm{T}]^{-1}$ and the number $\pi/4$ is dimensionless.

S32 *Could you summarize for me the use of dimensional analysis in physics?*

T32 For physicists, dimensional analysis is second nature, so to speak—it is used so much that one is not even conscious of it. First, it provides a ready check on the equations. A result like

$$v = \sqrt{T\mu}$$

(where v is the speed of transverse waves on a wire of linear density μ under tension T) can be discarded without going through its 'derivation' because the dimensions of the two sides do not match. (The correct formula is $v = \sqrt{T/\mu}$.) Similarly if in an expression we see factors like $\sin(\omega x)$, where ω is an angular frequency and x is the distance, we can be sure that an error has crept in. Of course, dimensional analysis can never tell you that an equation is correct; but it can point out that a certain equation is definitely wrong.

Second, it helps you recall formulae and equations in physics. For example, when I try to write down the transformation equation for time in special relativity, I often linger over the second term in the numerator:

$$t' = \frac{t-?}{\sqrt{1-\frac{v^2}{c^2}}}$$

I do know the term contains x. Also, I know that it should go to zero as v/c goes to zero. Now, I start looking at the dimensions. The denominator has 1 and v/c, both dimensionless. That is fine. The term I am looking for has the dimension of time. Aha! it is probably xv/c^2. Yes, that checks with my memory too.

In the same way, if I have forgotten the units of some parameter, I just have to look at an equation containing that parameter to get it back. For example, the units of the gas constant R can be obtained by looking at the gas law equation $PV = nRT$. Since pressure is force divided by area, PV has units of energy. R, therefore, must have the units J mol^{-1} K^{-1}.

S33 *These seem to me rather trivial uses, though I agree that they could be important in practice. Is that all to dimensional analysis?*

T33 No Often it goes beyond the mere checking of dimensional consistency or deciphering dimensions of quantities in an equation. For example, if the derivation of a result in physics is long and involved, it may help us to see if the important features of the result can be obtained without going through the derivation, by invoking dimensional arguments only. An example quoted in many books is the formula for a viscous force F on a sphere of radius r moving with speed v in a medium of viscosity η. We start by writing

$$F = (r)^\alpha (v)^\beta (\eta)^\gamma.$$

Now if you go back to the dimensions of η, you can easily see that its dimensions are $[M][L]^{-1}[T]^{-1}$. Writing the dimensions of each quantity in the equation, we get

$$[M]^1[L]^1[T]^{-2} \sim [M]^\gamma [L]^{(\alpha+\beta-\gamma)}[T]^{-(\beta+\gamma)}$$

Finally, equating the powers of $[M]$, $[L]$ and $[T]$ we obtain

$$\alpha = 1,\ \beta = 1,\ \gamma = 1.$$

Therefore $\quad F \sim \eta r \nu.$

The proper derivation gives

$$F = 6\pi\eta r\nu.$$

Of course, you can never obtain a dimensionless constant like 6π from dimensional analysis. But it is nice to see that the dependence on the variables that you have in the correct formula has been predicted on the basis of dimensional considerations alone.

S34 *I have a doubt here. How do you know that one should take only r, v, η on the right side above? Why not take the density ρ of the sphere, its acceleration a etc.?*

T34 That is a very pertinent question. A basic limitation of this game of trying to 'derive' formulae from dimensional analysis is that you can never logically arrive at what the relevant variables are in a problem. To guess the relevant variables, you must use your intuition, experimental information or whatever. So what we have done above is not really any alternative derivation of the viscous force formula. Rather, given that r, v, η determine F, we see that dimensional considerations alone are sufficient to determine the nature of dependence of F on these variables.

But in my view the real power of dimensional analysis comes into play when we go to a new, unfinished part of physics.

S35 *I do not quite appreciate that. Please elaborate.*

T35 Well, in an unknown domain, correct equations of physics cannot be 'served' to you. You need to discover them. Now, discovery is often a matter of playing a game—trying out ideas and hunches which may not be quite clear to the 'player' himself. Dimensional analysis is a real handy toy for this game.

S36 *That sounds very interesting. Any examples?*

T36 I shall give you two examples. First, let us imagine what Niels Bohr might have been thinking in early 1912 when he was at Manchester. Now you may have learnt that Bohr wanted to explain Balmer series in the hydrogen the spectrum, and eventually his model did explain it. That, no doubt, is true. But you must also know that Bohr came to learn of Balmer's formula only in early 1913. The atomic problem was worrying him much earlier.

S37 *What was worrying Bohr about the atomic problem?*

T37 The stability of the Rutherford atom was one problem. But Bohr also wondered as to what determined the size of an atom. You see there was already evidence from many sources that atoms are not point objects but have sizes of the order of 10^{-8} cm.

S38 *So what if the size is 10^{-8} cm?*

T38 There is no way in the Rutherford model to arrive at this size out. Classically, an electron can revolve around a nucleus at any distance, small or big. Bohr was puzzled by the fact that there seemed to be a minimum to the orbital size.

S39 *What has that to do with dimensional analysis?*

T39 A lot. Remember Bohr was exploring an unknown domain. He must have 'played' with various ideas to explain this size. And perhaps one of the ideas was based on dimensional analysis.†

S40 *I am thrilled. Please continue.*

T40 Anybody's first guess would be that the constants e (electronic charge) and m_e, (mass of an electron) would appear in the eventual formula for atomic size. (A little reflection shows that the nuclear mass, being much greater than m_e, will at most provide small correction to the atomic size. Nuclear charge Z, however, is important. In what follows we take $Z = 1$ (hydrogen atom)). You cannot, however, get a quantity of the dimension of length from these two constants. So you should look for yet another constant. Now the speed of light c was, by then, known to be a fundamental constant of nature. Using e, m_e and c it is easy to see that the quantity e^2/m_ec^2 does have the dimension of length. Still that does not solve the problem. (Note: e^2 should be replaced by $e^2/4\pi\epsilon_0$ in SI)

S41 *Why not?*

T41 When you substitute the values of e, m_e and c you find e^2/m_ec^2 to be 2.8×10^{-15} m which is far too small compared to the scale

† The actual historical process leading up to the discovery by Bohr of the atomic model was far more involved than has been suggested in the following paras. Our interest here, however, is not to describe the history of this discovery but to illustrate the exploratory use of dimensional analysis.

of atomic size. No, this quantity is not what we are looking for. On second thoughts, it is perhaps not surprising that the attempt above has not worked. The constant c has a role to play in the relativistic domain. That is not the typical domain of electronic motion in atoms. We know that electronic energies in atoms are of the order of eV to keV. The rest mass of an electron is about 0.5 MeV. This suggests that atomic size should be mainly determined by non-relativistic considerations. So we discard the constant c from the game.

S42 *Well, then we are back to e and m_e. What might be Bohr's thoughts at this stage?*

T42 Bohr obviously needed another constant. He was aware of Planck's constant h which determined the minimum of energy (for a given frequency) that can be exchanged between matter and radiation. By means of marvellous intuition, he guessed that the minimum in orbital size of an electron in an atom might not be completely unrelated to this, i.e. h might be a relevant constant for atoms too.

S43 *That is just a guess. What could be a clue to this guess, apart from intuition?*

T43 The clue might have come from dimensional analysis. Try to construct a quantity with the dimension of length using e, m_e, and h. You will discover that the combination $h^2/m_e e^2$ has the dimension of length. What is more, putting in the values of e, m_e, and h we get the length to be 2×10^{-9} m which is close to the order of the atomic size. (Actually if you put $\hbar$ in place of h you get 0.53×10^{-10} m—almost exactly the size of the hydrogen atom!)

S44 *That looks like mere manipulation with constants. It is hardly a theory of atomic size.*

T44 You are right. At this stage Bohr was far from any theory. Yet he got the first important idea, namely that Planck's constant h was likely to appear in any theory of the atom. The rest is history.

S45 *What is your second example of the use of dimensional analysis in exploring unknown domains of physics?*

T45 This relates to a curious, so far unproved, conjecture of the great physicist P A M Dirac. Now while in the first example we have only speculated about Bohr's thinking, Dirac, we know, actually played the game of dimensional analysis with fundamental constants of nature.

S46 *What is Dirac's conjecture?*

T46 Dirac obtained two dimensionless numbers from these constants, namely $e^2/(4\pi\epsilon_0 G m_e m_p)$ and $(4\pi\epsilon_0 m_e c^2 cT/e^2)$. The first is the ratio of the electrostatic force and the gravitational force between an electron and a proton. The second number represents the ratio of the radius of the universe (the present age of the universe T, which is estimated to be about ten billion years, multiplied by c, the velocity of light) and the so-called classical radius of the electron. When numerical values of the constants are substituted in these expressions the numbers turn out to be very large ($\sim 10^{40}$) and nearly equal (within an order of magnitude). Dirac was struck by the near coincidence of these two large numbers. He hypothesized that in some future theory they should turn out to be equal:

$$\frac{e^2}{4\pi\epsilon_0 G m_e m_p} = \frac{4\pi\epsilon_0 m_e c^2 cT}{e^2}.$$

S47 *That seems very speculative. In any case what is the implication of this equality?*

T47 T in the above equation, recall, stands for the age of the universe. So, obviously, it is not constant in time. T increases with the age of the universe. Consequently, if Dirac's conjecture is true, one or more of the other 'constants' in the equality must change with time. People have conjectured that the gravitational constant G might have (extremely weak) dependence on time. So far the conjecture remains unproved.

S48 *What is the moral of the story?*

T48 Try to make dimensional analysis part of your instinctive response to any problem in physics. For any physical situation, look at the relevant quantities, their dimensions and orders of magnitudes; note any significant dimensionless ratios and understand the approximations involved in terms of these ratios. And, of course, do check the dimensional consistency of your equations. When the 'territory' is unknown, use dimensional analysis as one of the tools to make progress.

S49 *I am already a convert. Shall we move on?*

2

Scalars and Vectors

Vectors add differently from scalars

S1 *I know the distinction between scalars and vectors: a scalar is a quantity that has only magnitude but no direction; a vector is a quantity that has magnitude as well as direction. What is so important about this distinction?*

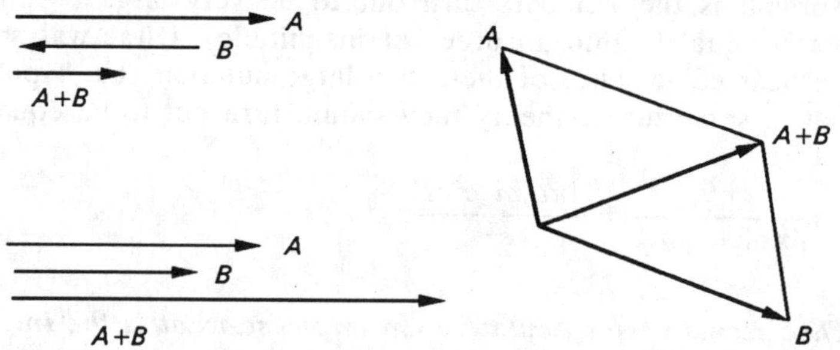

Fig. 2.1 The parallelogram law of addition of vectors ***A*** and ***B*** leads to resultants of different magnitudes and directions, depending on the angle between them.

Tl One important distinction is that vectors 'add' differently from scalars. Scalars add according to the usual rules of arithmetic, whereas vectors add according to the parallelogram law. This makes all the difference in so many situations in which we have to 'add' up physical quantities to obtain the resultant. For example, a body which has received energy of 12 J from one source and 9 J from another has received a total of 21 J of energy. But a body subject to two forces of 12 N and 9 N may have a resultant force anywhere between 3 N and 21 N, depending on the angle between the two forces. (see Fig. 2.1)

S2 *A scalar has magnitude only. Does this mean a scalar can never take negative values?*

T2 No, there is no such restriction in general. Scalars like mass, speed and kinetic energy, are, of course, non-negative. But a quantity like potential energy is also a scalar and it can take any value, positive or negative. A scalar should not be associated with any direction in space, that is all. An electric charge is also a scalar because it is not associated with any direction in space. It can take any value, positive, negative or zero.

S3 *But sometimes positiveness or negativeness of a quantity itself can denote direction. For example, on a straight road velocity along one direction may be taken to be positive, and, in the opposite direction, negative.*

T3 That is a correct observation. The sign of a physical quantity may or may not have to do with direction. It depends on what quantity we are talking about. The sign of charge or potential energy does not have anything to do with direction. On the other hand the sign of displacement, velocity or force ***in one dimension*** is related to the direction of these quantities.

S4 *Why do you only say that about one-dimensional vectors only?*

T4 Because the directions of vector quantities in two dimensions (vectors in a plane) or of vectors in three-dimensional space cannot be characterized by merely assigning a sign to the numbers that denote their magnitudes. In one dimension (vectors confined to a line), however, there are only two possibilities of direction which can be taken care of by the algebraic sign + or −.

S5 *Does that mean there is no difference between scalars and one-dimensional vectors?*

T5 None mathematically. Algebra of one-dimensional vectors (which means the rules of operations on these quantities) is the same as that of scalars. For example, the addition of two charges $-5C$ and $+2C$ is $-3C$. In the same way the addition of two forces − 5 N and + 2 N along the same line gives − 3 N. The negative sign in $-3C$ has no relation to any direction. The negative sign in − 3 N, however, gives the direction of the force on the given line.

S6 *Are there situations where one-dimensional vectors are relevant?*

T6 Consider a ball thrown vertically upwards. Here velocity, acceleration, displacement and force are all one-dimensional vectors.

Box 2.1

Consider, for simplicity, vectors in two dimensions. The resultant $\boldsymbol{C}$ of vectors $\boldsymbol{A}$ and $\boldsymbol{B}$ is given by

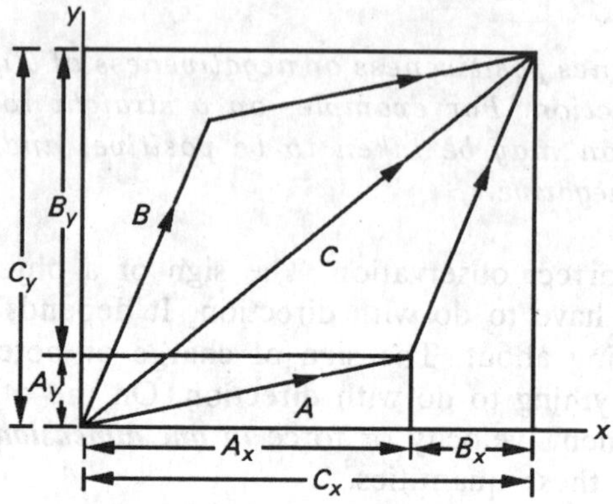

Fig. 2.2 Vector addition rule in terms of components.

the parallelogram law as shown in Fig. 2.2. We choose rectangular cartesian axes as shown in the figure. The components of the three vectors, along these axes, are (A_x, A_y), (B_x, B_y) and (C_x, C_y). It is quite evident from the figure that $C_x = A_x + B_x$ and $C_y = A_y + B_y$. Thus the law of addition of pairs follows from the parallelogram law. Its generalization to three dimensions is straightforward.

S7 *In what way is the algebra of vectors in two and three dimensions different from the algebra of scalars?*

T7 We have seen that already in *T*1. But I shall repeat the point because it is very crucial. Scalars add like numbers in arithmetic. Vectors add according to the parallelogram law. If $\boldsymbol{C} = \boldsymbol{A} + \boldsymbol{B}$, then C, i.e. the magnitude of $\boldsymbol{C}$, is not equal to $A + B$ but is given by

$$C = \sqrt{A^2 + B^2 + 2AB\cos\theta}$$

where θ is the angle between $\boldsymbol{A}$ and $\boldsymbol{B}$. Only for $\theta = 0$, $C = A + B$.

Another way of seeing the difference between scalars and vectors is as follows. A scalar is associated with one real number, a vector in a plane is associated with an ordered pair of numbers. In the same way, a vector in three dimensions can be associated with an ordered triplet of numbers. To add vectors, all one does is arithmetically add the corresponding numbers in the pairs or the triplets. For example, if vector $\boldsymbol{A}$ is associated with a pair $(3, -1)$ and vector $\boldsymbol{B}$ with the pair $(5, 3)$; then the pair associated with $\boldsymbol{A}+\boldsymbol{B}$ is $(8, 2)$.

S8 *These seem to be ordinary rules of addition. What is the difference?*

T8 That is a serious error. The rule regarding addition of pairs (or triplets) given above does ***not*** follow from the arithmetic of numbers. It follows from the parallelogram law of vector addition.

S9 *Is time a scalar or a vector?*

T9 Ofcourse, a scalar.

S10 *Why, but time also has a direction. Present to the past is time backwards and the present to the future is time forwards.*

T10 Be careful. You are using the word 'direction' in some kind of general sense. When we talk of a vector, we are talking of a direction in space. Time is not associated with any direction in space, so it is a scalar.

Free and localized vectors

S11 *That is clear. Let me turn to something else. I see in books that vectors are freely displaced from one point to another (keeping their length and direction constant). Is a vector not supposed to be located at a particular point in space?*

T11 Ordinarily, in vector algebra, vectors are defined by their magnitude and direction only, and have no association with any particular location in space. Two vectors are equal if both have the same magnitude and direction, regardless of their location. Therefore, for vector operations (addition, subtraction, etc.), there is no error in displacing a vector from one point to another provided you do not change its length or direction.

S12 *I am still not clear. A force, for example, is a vector quantity. But obviously the effect of a force depends not only on its magnitude and direction but also on what point it is applied.*

T12 You are right. For instance, the rotational effect of a force on a rigid body depends on its line of application, because magnitude of torque = magnitude of force × lever arm, and the latter depends on the line of application of the force (see Fig. 2.3).

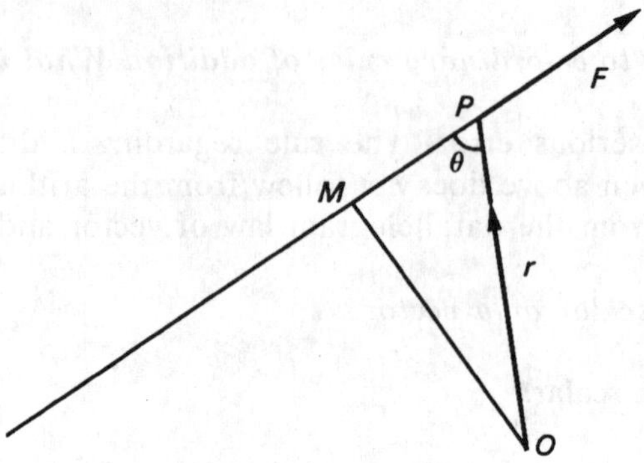

Fig. 2.3 The lever arm of the force $\boldsymbol{F}$ acting at P is $OM = r \sin\theta$.

S13 *Why do you use the words 'line of application'? Why don't you merely refer to the point where the force is applied.*

T13 Because the particular point on the line where the force may be applied is immaterial. The lever arm is simply the distance of the line from the point O around which the torque is to be calculated.

S14 *Is it generally true in mechanics that the effect of a given force depends on its line of application and not on the particular point where it is applied?*

T14 No. That is true if you are considering rigid bodies. For a body that is elastic or deformable, the effect of a force also depends on the point where it is applied. For example, in Fig. 2.4, two forces of equal magnitude and opposite direction, acting along the same line of application, are exerted on an elastic rod. If the effect of a force depended only on its line of application, the resultant would be a null force with no effect. But that, clearly, is not true; the length of the rod is greater when the forces act. Thus, in this case, the effect of a force is determined not solely by its line of application, but by its point of application as well.

Fig. 2.4 Two equal and opposite forces applied at the two ends of an elastic rod do not produce a null effect.

S15 *Then you agree with what I said in S12. It means I cannot displace vectors like the way it is shown in books.*

T15 Vectors which have no association with any particular location in space are called 'free vectors'. Free vectors can always be displaced.

S16 *Oh, that is begging my question. It just means that force, for example, is not a free vector and, therefore, vectors associated with forces cannot be displaced.*

T16 You are right. Some people like to call a vector which is bound to a particular location in space as a 'localized' or 'bound' vector. Force is an example of a localized vector.

S17 *Still, it is customary in books to show forces on a body displaced so that they all appear together at one point, and then use the parallelogram law of addition. But you just agreed that force is a localized vector and cannot be displaced.*

T17 You should always understand that certain things which may not be true in general could be correct in limited contexts. Consider, as you have said, a body under the action of a number of external forces applied at different points. Now if we are interested only in the translational motion of the body as a whole (i.e., the motion of its centre of mass), then it is perfectly correct to displace the forces such that they are at a point (centre of mass) and add them up by the usual laws of vector addition. In many elementary problems in mechanics, it is the translational motion or translational equilibrium that we investigate. In that context, it is quite correct to treat forces as free vectors. Of course, the moment you are interested in say rotational motion or rotational equilibrium, you cannot displace forces; and in that context force is not a free vector.

S18 *Can you give me some other example of a localized vector?*

T18 There are many examples. Suppose we are given a charge distribution in space. With each point in space, you can associate a vector, namely the force a unit positive charge would experience if placed at that point. This collection of vectors is called an electric field. Likewise one can also define a magnetic field. Since vectors here are associated with points in space, the vectors representing the electric or magnetic fields are clearly localized vectors.

S19 *Is there a similar notion for scalars also? Can scalars also be localized? Are there scalar fields?*

T19 Yes, of course. Inherent in the idea of a field is the association of something with each point in space. If that something is a scalar quantity, then we have a scalar field. If you associate a vector with each point, then we have a vector field. Electric and magnetic fields are instances of vector fields. In an environment where temperature may vary from point to point, I can talk of a temperature field, which is a scalar field. Likewise, we talk of potential (gravitational or electric) as a scalar field.

Multiplication of two vectors

S20 *I now understand the notion of localized vectors. Let me ask something else. For scalars, we have only one type of multiplication. How is it that we have two types of multiplication for vectors?*

T20 Look at it this way. 'Addition' or 'multiplication' are merely names given to binary operations on a set of objects. (Binary operation means an operation involving two elements of the set.) For scalars, i.e. ordinary numbers, the meanings of these operations are familiar to us. Now, vectors are a different set of objects from scalars. For this set, the meanings of 'addition' and 'multiplication' have to be defined afresh. These definitions do not follow from the corresponding definitions for scalars. You already know that the 'addition' of vectors is defined by the parallelogram law. In the same way, we need to define the meaning of multiplication of two vectors.

S21 *O.K. but why two different types of multiplication?*

T21 It so happens that besides 'addition' it is possible to have two very useful binary operations on vectors. Do not worry that we will use the word 'multiplication' for these operations. What we call these operations is not important. The significant thing is that two useful binary operations can be defined.

S22 *I do not see why you use words like 'useful definition'. A definition is a definition. It is either correct or incorrect. Either vectors have two types of multiplication, or they do not.*

T22 There is an important issue here and I believe you need a different orientation in your thinking. Otherwise, at many points in mathematics and physics, you will ask similar questions and find yourself puzzled. I repeat, the so-called scalar multiplication and vector multiplication are two different and useful definitions of binary operation on vectors (besides addition).

S23 *I still do not get you.*

T23 You must understand that no definition is right or wrong, because a definition is our construct, our choice. But, of course, a definition may be very significant and useful, or it may be trite and useless. The following two binary operations have been found to be useful. From two vectors **A** and **B**, I can construct a scalar $AB\cos\theta$ where A, B are magnitudes of **A** and **B**, and θ is the angle between **A** and **B**. Likewise I can construct a vector whose magnitude is $|AB\sin\theta|$ and whose direction is normal to the plane of **A** and **B**. (The sense of the normal is determined by the motion of the right-handed screw rotated from **A** to **B** through the smaller angle. (see Fig. 2.5)

(Note, each of these objects involves a product of magnitudes of **A** and **B** besides other things. So I call *ABcosθ* as a scalar product of **A** and **B** and the vector defined above as vector product of **A** and **B**.)

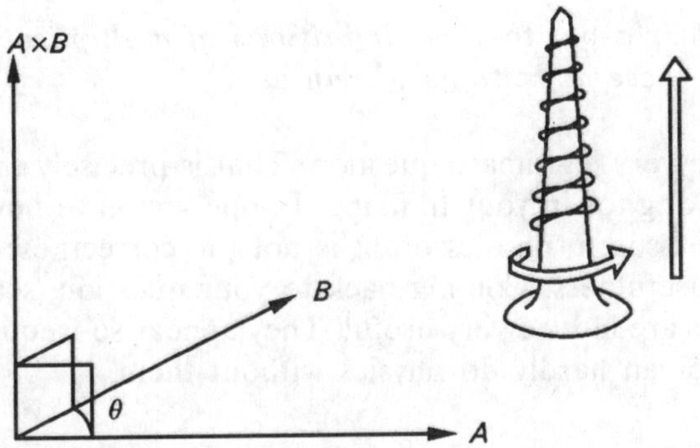

Fig. 2.5 Definition of the vector product of two vectors.

S24 *What do you mean by the smaller angle between* **A** *and* **B** *in your definition of the vector product?*

T24 The angle between any two directions in space can be given two values: θ and $2\pi - \theta$. Of these, one will be smaller than (or equal to) π, and the other will be greater than (or equal to) π. In the scalar $AB\cos\theta$ it is immaterial which value I take because $\cos(2\pi - \theta) = \cos\theta$. But in specifying the direction of the vector product of $\boldsymbol{A}$ and $\boldsymbol{B}$ I must choose one of these angles, because the directions of advance of a right-handed screw rotated through the two angles are opposite. We choose the smaller angle.

S25 *Well then, where do we stand now?*

T25 We started out trying to define 'multiplication' of two vectors. We have seen that from two vectors we can construct both a scalar and a vector. Next, for convenience, I give a notation for these operations. I denote the object named scalar product by the symbol $\boldsymbol{A} \cdot \boldsymbol{B}$. Similarly, I denote the object named vector product by the symbol $\boldsymbol{A} \times \boldsymbol{B}$. Therefore

$$\boldsymbol{A} \cdot \boldsymbol{B} = AB\cos\theta\,,$$

$$\boldsymbol{A} \times \boldsymbol{B} = AB\sin\theta\ \widehat{n}.$$

where $\widehat{n}$ is a unit vector in the direction of $\boldsymbol{A} \times \boldsymbol{B}$ as specified in *T*23. Sometimes the notations themselves generate new names. Thus the scalar product $\boldsymbol{A} \cdot \boldsymbol{B}$ is also called a dot product of $\boldsymbol{A}$ and $\boldsymbol{B}$, and the vector product $\boldsymbol{A} \times \boldsymbol{B}$ is also called the cross product of $\boldsymbol{A}$ and $\boldsymbol{B}$.

S26 *I am clear about the two definitions of multiplication of vectors, but are these definitions of any use?*

T26 This is a very legitimate question. That is precisely the orientation I was looking for in your thinking. I hope you have now got the point that the issue in physics often is not the correctness of a definition but its usefulness. Coming back to your question, scalar and vector products are indeed very useful. They appear so frequently in physics that one can hardly do physics without them.

S27 *Can you give me some examples of their usefulness?*

T27 Consider a force $\boldsymbol{F}$ applied over a displacement $\boldsymbol{d}$ of a particle. The scalar built out of $\boldsymbol{F}$ and $\boldsymbol{d}$, i.e., $\boldsymbol{F} \cdot \boldsymbol{d}$ has important physical significance. It denotes the work done by $\boldsymbol{F}$ resulting in a change in kinetic energy of the particle. In the same way, the vector built from the position vector $\boldsymbol{r}$ and momentum $\boldsymbol{p}$ of a particle, namely the cross product $\boldsymbol{r} \times \boldsymbol{p}$, is an important quantity, the angular momentum $\boldsymbol{L}$. The vector product $\boldsymbol{\tau} = \boldsymbol{r} \times \boldsymbol{F}$ denotes the torque produced by the force $\boldsymbol{F}$ around the origin. Both $\boldsymbol{L}$ and $\boldsymbol{\tau}$ have great significance in physics, especially in problems of rotational motion. In electromagnetism, the vector $\boldsymbol{E} \times \boldsymbol{B}$ is related to the radiative energy flow. The volume of a parallelepiped of sides $\boldsymbol{A}$, $\boldsymbol{B}$ and $\boldsymbol{C}$ is given by $\boldsymbol{A} \cdot (\boldsymbol{B} \times \boldsymbol{C})$. The examples abound.

Tensors

S28 *I grant you that these definitions of multiplication of vectors are useful. But since, as you said, definitions are our constructs, can't we have some more definitions of multiplication? From the 'multiplication' of two vectors* $\boldsymbol{A}$ *and* $\boldsymbol{B}$*, why do I necessarily have to get a scalar or a vector? Can't I get something else?*

T28 It is for you to try and see if you can arrive at any additional useful definitions. But I know at least one more way of 'multiplying' two vectors. From a vector $\boldsymbol{A}$ with components A_i, $i = x, y, z$ and vector $\boldsymbol{B}$ with components B_j, $j = x, y, z$ I can construct nine product quantities A_iB_j. These nine quantities together form an object which is neither a scalar nor a vector.

S29 *What then is this object?*

T29 This 9-component object is called a tensor (of rank 2). We can denote it by T_{ij} (where $i = x, y, z$; $j = x, y, z$) and identify its components as $T_{xx} = A_xB_x$, $T_{xy} = A_xB_y$, etc.

S30 *But* A_i, B_j *are components of vectors* $\boldsymbol{A}$, $\boldsymbol{B}$. T_{ij} *are components of what?*

T30 One usually works with tensor components only. But if you insist, I can introduce an object $\overleftrightarrow{\mathbf{T}}$, whose components are T_{ij}. Such an object is called a dyadic. A dyadic is a sort of 'double vector'. To get its components, simply take the appropriate scalar product from either side. Thus

$$T_{xy} = \hat{i} \cdot \overleftrightarrow{\mathbf{T}} \cdot \hat{j},\ T_{yz} = \hat{j} \cdot \overleftrightarrow{\mathbf{T}} \cdot \hat{k}, \text{ etc.}$$

where $\hat{i}$, $\hat{j}$ and $\hat{k}$ are unit vectors along x, y and z axes, respectively. The dyadic constructed out of vectors $\boldsymbol{A}$ and $\boldsymbol{B}$ may be denoted by $\boldsymbol{AB}$. Its components are

$$A_xB_x = \hat{i} \cdot \boldsymbol{A}\,\boldsymbol{B} \cdot \hat{i}, \qquad A_yB_z = \hat{j} \cdot \boldsymbol{A}\,\boldsymbol{B} \cdot \hat{k}$$

and so on.

S31 *Another question. We have the 'addition', 'subtraction' and 'multiplication' of two vectors. Is there any meaning to the division of* $\boldsymbol{A}$ *by* $\boldsymbol{B}$*?*

T31 If $\boldsymbol{A}$ and $\boldsymbol{B}$ are in the same (or opposite) direction, i.e., $\boldsymbol{A}$ is a scalar multiple of $\boldsymbol{B}(\boldsymbol{A} = \lambda\boldsymbol{B})$, then it is natural to define division of $\boldsymbol{A}$ by $\boldsymbol{B}$ as

$$\frac{\boldsymbol{A}}{\boldsymbol{B}} = \lambda$$

But for any two arbitrary vectors $\boldsymbol{A}$ and $\boldsymbol{B}$ it does not appear possible to give a straightforward definition of division. However, if one means by division of $\boldsymbol{A}$ by $\boldsymbol{B}$ as 'something' which multiplied by $\boldsymbol{B}$ gives you $\boldsymbol{A}$, then some progress is possible. If $\boldsymbol{A}$ and $\boldsymbol{B}$ are related by an expression of the type

$$\boldsymbol{A} = \overleftrightarrow{\mathrm{T}} \cdot \boldsymbol{B},$$

where $\overleftrightarrow{\mathrm{T}}$ is a 'double vector', a dyadic, then, in a sense, $\overleftrightarrow{\mathrm{T}}$ is what you get when $\boldsymbol{A}$ is 'divided' by $\boldsymbol{B}$. In component form, this equation looks like

$$A_i = \sum_j T_{ij} B_j \ , \ i = x, \ y, \ z,$$

which shows that our quest for trying to define $\boldsymbol{A}$ divided by $\boldsymbol{B}$ does not quite give us any simple answer; rather it ends up in the notion of tensors.

S32 *We seem to be ending up in these objects called tensors, both while trying to generate new types of multiplication of two vectors and also in our attempt to define division of two vectors. But in the same spirit that you emphasized in T26, is the tensor* T_{ij}*, the way you have defined, a useful construct at all?*

T32 It is indeed so. Tensor analysis is one of the most useful branches of mathematics which now underlies basic physical theories. It is especially important when in a theory issues such as, how quantities behave under such and such transformations, which quantities are invariant, whether the given physical laws are the same for different frames of reference, etc. are in focus. But since these matters are somewhat more advanced, I will not dwell on them here.

S33 *Still you can give me some feel about these objects. Are there physical quantities which are tensors?*

T33 Yes, there are numerous examples of tensors. You will realize from the preceding discussion that whenever in a physical situation a vector quantity $\boldsymbol{A}$ is related to another vector quantity $\boldsymbol{B}$ that is not necessarily in the same direction as $\boldsymbol{A}$, the 'coefficients' connecting $\boldsymbol{A}$ and $\boldsymbol{B}$ are going to be tensors.

S34 *Are there situations of that kind?*

T34 Yes. Consider a dielectric placed in an electrostatic field $\boldsymbol{E}$. The field will produce a polarization $\boldsymbol{P}$ in the medium. Now, if the direction of $\boldsymbol{P}$ is the same as that of $\boldsymbol{E}$, we may write

$$\boldsymbol{P} = \chi \boldsymbol{E}$$

where χ is a scalar called the electric susceptibility of the medium. Such a relation is true only for isotropic dielectrics. Some dielectrics (for example, anisotropic crystals), however, exhibit significant direction dependence and in such cases, $\boldsymbol{P}$ is not necessarily in the same direction as $\boldsymbol{E}$. For such anisotropic dielectrics, we can write

$$P_i = \sum_j \chi_{ij} \, E_j, \qquad i = x, y, z$$

where χ_{ij} is a 9-component object, a tensor of rank 2. It is known as the susceptibility tensor.

Other examples of tensors are the moment of inertia tensor, stress tensor in the theory of elasticity, electromagnetic field strength tensor, etc.

Vector nature of area

S35 *Perhaps we have had enough of tensors. Let me ask another little doubt. Is area a scalar or a vector?*

T35 The magnitude of area is, of course, a scalar quantity. But any element of area is not fully specified by its magnitude alone. We must know how that area element is oriented. For example, a 1 cm^2 area in x-y plane and another 1 cm^2 area in y-z plane are not identical things (see Fig. 2.6). They have the same magnitude, but are oriented differently. So an area element is specified fully only if you give its magnitude as well as its direction; it is a therefore vector quantity.

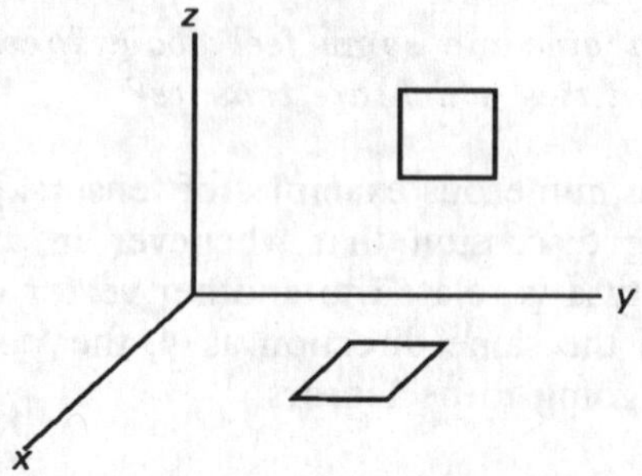

Fig. 2.6 Area elements with different orientations.

S36 *But how do we specify the 'direction' of an area element? On the rectangle shown in Fig. 2.6, I can draw lines in all kinds of directions. Which of these shows the direction of area?*

T36 None. The direction of area element is the direction from which you can tell the orientation of the area. Obviously, the orientation is specified if you give the direction of the normal to the area element.

S37 *I note you use the words 'area element' by which I guess you mean a small, strictly infinitesimal, area. What is the point?*

T37 If we are dealing with a finite area in a plane, I can associate with it a definite direction normal to its plane. For a finite non-planar area the direction of normal may differ from point to point. An infinitesimal area, however, can be treated as a planar area and, therefore, has a definite direction for its normal. This is why I have been talking of an area element and not a finite area.†

S38 *But even a planar area does not have a definite direction. The normal can be 'up' or 'down'. Which of the two alternatives does one choose?*

T38 You are right about the ambiguity regarding the sense of direction of the vector associated with an area element. There is no intrinsic way to resolve this ambiguity, but it can be handled through some agreed conventions. For an area element of a closed surface, by convention, we take the outward normal as the direction of the vector area element. But there is no 'outward' or 'inward' for an open surface bounded by some curve. In this case I can agree on a convention relating the direction of vector area element to the

† The vector area of a finite non-planar surface can, however, be obtained by integration over the surface.

sense of traversing the curve. Thus to decide the direction of an area element we choose a sense of traversing the curve bounding the element. Now if a right-handed screw is rotated in this sense, the advance of the screw gives the direction of the area element (see Fig. 2.7).

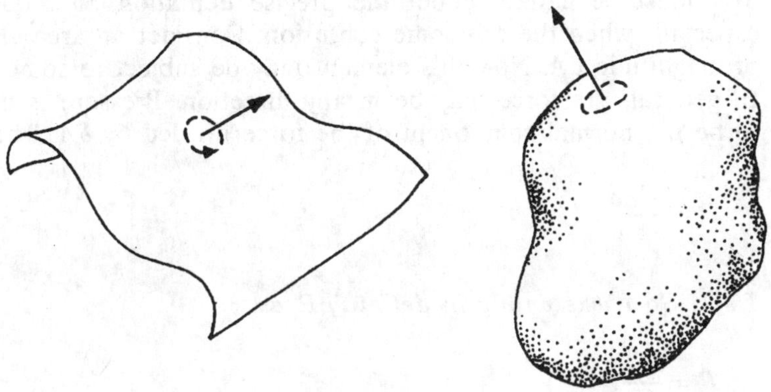

Fig. 2.7 Direction of area element of a closed and an open surface.

S39 *I understand the vector nature of an area element, but it all seems so unnecessary. Why not consider area simply as length times breadth, or something like that? Why must I bother about its direction?*

T39 It depends on what you are doing. If what matters to you is just the magnitude of the area, and not how it is oriented, then your object of interest is a scalar. (The magnitude of a vector is a scalar.) Most problems of school mathematics need only the magnitude of the area and that is why you have this lingering impression of area being a scalar. But in many problems, the orientation of the area is equally important. For example, in a running stream the amount of water flowing across an area element obviously depends on the orientation of the area. Similarly, if we want to know the flux of an electric field $\boldsymbol{E}$ passing through an element of area, both the magnitude and orientation of the area are important. I must then associate a vector $\delta \boldsymbol{A}$ with that element and write the flux as $\boldsymbol{E} \cdot \delta \boldsymbol{A}$.

Pressure and current are scalars

S40 *Let us consider another physical quantity : pressure. Is it a scalar or a vector?*

T40 A scalar obviously.

S41 *It is not so obvious to me. Pressure is supposed to be force divided by area. Now force is a vector, and area in this definition, I suppose, stands for the magnitude of area. A vector divided by a scalar should be a vector.*

T41 You must be careful about the precise definition of a quantity, especially when there is some confusion. Consider an area element of magnitude δA. Now this element may be subject to some force. In general, the force may be in any direction. Pressure is defined to be the normal component of the force divided by δA. That is

$$P = \frac{\boldsymbol{F} \cdot \hat{\boldsymbol{n}}}{\delta A}.$$

S42 *I see. So I was wrong in defining P as*

$$P = \frac{\boldsymbol{F}}{\delta A}.$$

*Well, the way you have defined, it is clear that pressure is a scalar. But just a little query. If **F** is not normal to the element, it will also have a tangential component. What happens to that?*

T42 That component produces another kind of stress, the so-called shearing stress. You see, a body under the action of a general deforming force experiences different kinds of stress, leading to different kinds of strain. The normal component of the force gives rise to 'volume stress' (i.e. pressure) leading to volume change but no change in shape. The tangential component produces shearing stress leading to changes in the shape of body.

S43 *I have just had an idea of how pressure might be a scalar. Pressure may be defined by*

$$P = \frac{\boldsymbol{F}}{\delta \boldsymbol{A}}.$$

The vector nature cancels in the numerator and denominator and hence P is a scalar. Am I right?

T43 You are completely wrong. Please go over what we discussed in $T31$ again. Division of $\boldsymbol{F}$ by $\delta \boldsymbol{A}$, if you insist on writing the way you have written, makes sense only if $\boldsymbol{F}$ and $\delta \boldsymbol{A}$ are in the same direction. If $\boldsymbol{F}$ and $\delta \boldsymbol{A}$ have different directions, your definition does not make sense. The words 'vector nature cancels' are particularly objectionable. Vector algebra has its own rules and you should not uncritically take over all the words and operations that you are used to in arithmetic.

S44 *Still, you did try to do something about the division of* ***A*** *by* ***B*** *when* ***A*** *and* ***B*** *are not in the same direction.*

T44 I never said there was anything like ***A/B*** (when ***A*** and ***B*** have different directions) but rather used the question to lead you up to tensors.

S45 *Here too then* ***F*** *is a vector and* $\delta \boldsymbol{A}$ *is also a vector not necessarily in the same direction. Would relating the components of* ***F*** *to the components of* $\delta \boldsymbol{A}$ *lead me to tensors?*

T45 Yes, it will. You can guess the equation yourself:

$$F_i = \sum_j P_{ij}\, \delta A_j,\ i = x,\ y,\ z.$$

This, very briefly, is the way one introduces the stress tensors P_{ij} in the theory of elasticity. The details cannot be gone into, but I think you have got the point.

S46 *Next, is electric current I a scalar or a vector quantity?*

T46 Let us see the definition of this quantity. First there is a quantity called current density. It is by definition a vector, denoted by ***j***. The direction of ***j*** is the direction of the flow of positive charges. The magnitude of ***j*** at a point is the amount of charge flowing per second through a small area normal to the flow around that point, divided by the magnitude of that small area. It is easy to see that

$$\boldsymbol{j} = \rho\, \boldsymbol{v}$$

where ρ is the charge density at the point, and ***v*** the velocity of the flow.

S47 *How is* ***j*** *related to I ?*

T47 The definition of a current always involves a surface, a cross-section of some conductor, etc. Current *I* across a surface is the total charge flowing across the surface per second. Now, note that at each point on the surface, ***j*** or the direction of flow may not be normal to the given surface. To get the amount of charge flowing across the surface, we must dot $\boldsymbol{j}$ with the normal $\widehat{\boldsymbol{n}}$ to the surface at the point. Further $\boldsymbol{j}$ refers to the rate of charge flow per unit area. There is no 'per unit area' in the definition of I. Therefore, we must integrate over the area of the surface, i.e.

$$I = \int \boldsymbol{j} \cdot \widehat{\boldsymbol{n}}\, da,$$

which is clearly a scalar.

S48 *I understand this from your definition. But still I can't get rid of the impression that there is a direction to current and so it is a vector.*

T48 To convince you that current is not a vector, let us recall Kirchhoff's first law for steady currents at a junction. Suppose at a junction there are two incoming currents I_1, and I_2, and an outgoing current I_3 . According to this law, $I_3 = I_1 + I_2$. Now this equality is true no matter what the relative directions of the wires carrying currents. This surely cannot be the case if currents were vector quantities. (If $\boldsymbol{C} = \boldsymbol{A} + \boldsymbol{B}$, the equality $C = A + B$ for their magnitudes is not true for arbitrary directions of $\boldsymbol{A}$ and $\boldsymbol{B}$; it is true only for the case when $\boldsymbol{A}$ and $\boldsymbol{B}$ are in the same direction.)

Rotations

S49 *Let me turn to another thing. If I can associate a magnitude and direction with some quantity, is it necessarily a vector?*

T49 Usually, but not always. For example, rotation is associated with a direction in space (the axis of rotation) and magnitude (the amount of angle of rotation about the axis). Yet rotation is not a vector.

S50 *Why not?*

T50 Because two rotations do not necessarily commute. That is, if R_1 is one rotation and R_2 is another rotation, and if by addition we mean one rotation followed by the other, then you can convince yourself, geometrically that in general they do not commute. Symbolically we may express it thus

$$R_1 \oplus R_2 \neq R_2 \oplus R_1.$$

(If R_1 and R_2 have a common axis of rotation then they do commute.)

S51 *So what if they do not commute?*

T51 Vector addition obeys the parallelogram law and hence is necessarily commutative. Since addition of rotations is not commutative, rotations are not vectors.

S52 *This seems to mean the familiar definition of a vector as given in S1 is not completely adequate. What then is the precise way of defining a vector?*

T52 To answer you, let me translate the property of having magnitude and direction in space in different terms. Suppose I consider a simple vector, the position vector $\boldsymbol{r}$ of a point P with reference to some origin O in a co-ordinate system. This vector has components x, y and z along the three axes. † Now what do you think will happen if I rotate the axes to some other orientations and call the components of $\boldsymbol{r}$ along the new axes by x', y' and z'? (see Fig. 2.8.)

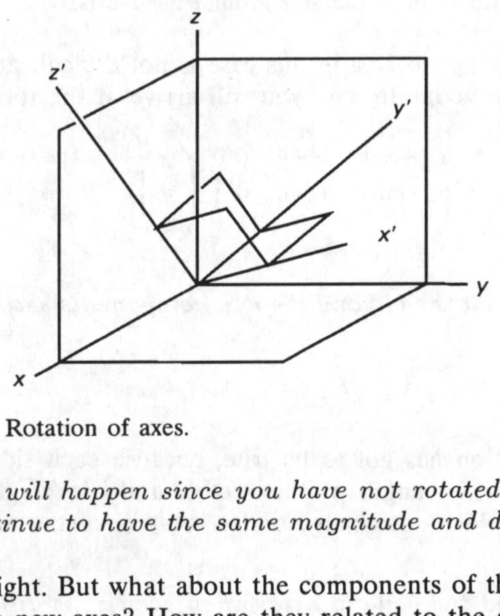

Fig. 2.8 Rotation of axes.

S53 *Nothing will happen since you have not rotated the vector* $\boldsymbol{r}$. *It will continue to have the same magnitude and direction.*

T53 That is right. But what about the components of this same vector $\boldsymbol{r}$ along the new axes? How are they related to the old components?

S54 *I do not know. It will depend on how the axes were rotated. I mean, by what angle and around which direction.*

T54 That is correct. But let me make life simple for you. Suppose we rotate the original axes around the z-direction by an angle θ, as

† We use rectangular Cartesian frames throughout our discussion.

shown in Fig. 2.9. What then is the relation between x, y, z and x', y', z'?

S55 *I do not know. All I can easily see is that* $z' = z$.

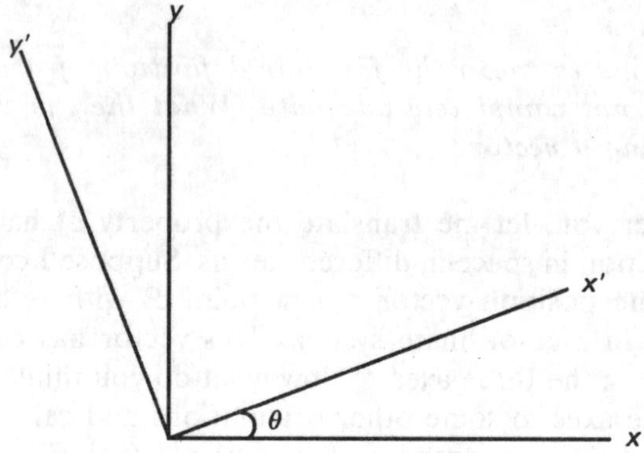

Fig. 2.9 Rotation of the axes around the z-axis.

T55 Relating x', y' to x, y in this case is not difficult at all. Try it as an exercise in geometry and you will arrive at the following result:

$$\begin{pmatrix} x' \\ y' \\ z' \end{pmatrix} = \begin{pmatrix} \cos\theta & \sin\theta & 0 \\ -\sin\theta & \cos\theta & 0 \\ 0 & 0 & 1 \end{pmatrix} \begin{pmatrix} x \\ y \\ z \end{pmatrix}.$$

S56 *I notice that the old and the new components satisfy the following relation:*

$$x'^2 + y'^2 + z'^2 = x^2 + y^2 + z^2.$$

T56 That relation has got to be true, because each side represents the square of the length of the vector, and the length of the vector obviously does not change under rotation of the axes.

S57 *How about the general case, when even* z' *is different from* z?

T57 For rotation in a plane, we just require one angle to specify the rotation. For a general rotation in space, one requires three angles, say α, β and γ. The transformation equations in this case are

$$\begin{pmatrix} x' \\ y' \\ z' \end{pmatrix} = \begin{pmatrix} R_{11} & R_{12} & R_{13} \\ R_{21} & R_{22} & R_{23} \\ R_{31} & R_{32} & R_{33} \end{pmatrix} \begin{pmatrix} x \\ y \\ z \end{pmatrix}.$$

where the coefficients R_{ij} are functions of α, β, and γ. They are a little complicated so I will not write them down here. But, of course, whatever they may be, the relation given in *S*56 has got to be true for the same reason as mentioned in *T*56.

Definition of a vector

S58 *I understand how the components of a position vector* **r** *transform as we rotate a frame to a new orientation. But why are you telling me all this? I asked you a straight question: What is the precise definition of a vector?*

T58 I am trying to answer just that question. A vector is a three-component object whose components transform under rotations of co-ordinate frames precisely like the components of position vector $\boldsymbol{r}$.

S59 *Can you repeat that?*

T59 $\boldsymbol{A}$ is a vector if its components A_x, A_y and A_z in one frame are related to its components A'_x, A'_y and A'_z in a rotated frame in exactly the same manner as in *T*57, i.e.

$$\begin{pmatrix} A'_x \\ A'_y \\ A'_z \end{pmatrix} = \begin{pmatrix} R_{11} & R_{12} & R_{13} \\ R_{21} & R_{22} & R_{23} \\ R_{31} & R_{32} & R_{33} \end{pmatrix} \begin{pmatrix} A_x \\ A_y \\ A_z \end{pmatrix}.$$

When I say the components of $\boldsymbol{A}$ transform in the same manner as the components of $\boldsymbol{r}$, I mean the transformation matrix here is identical to that appearing in *T*57.

S60 *That is clear. But what about the commutativity of vector addition? Does it follow from your definition?*

T60 No. Commutativity follows from the parallelogram law of vector addition, which is a separate requirement. We require that

$$(\boldsymbol{A}+\boldsymbol{B})_i = A_i + B_i.$$

This ensures that if A_i, and B_i transform under rotations of the axes as given in *T*59, $(\boldsymbol{A}+\boldsymbol{B})_i$, will also transform similarly.

S61 *Your technical definitions are clear to me. I am also clear that rotation cannot be associated with a vector because two rotations, in general, do not commute. But I have seen books using vector notation for angular velocity* $\boldsymbol{\omega}$. *I have also come across the vector relation* $\mathbf{v} = \boldsymbol{\omega} \times \mathbf{r}$. *How come a rotation (written symbolically as* θ*) is not a vector but* $\boldsymbol{\omega}$ *is a vector?*

T61 Look at it this way:

$$\boldsymbol{\omega} = \lim_{\delta t \to 0} \frac{\delta \theta}{\delta t},$$

i.e. in the definition of $\boldsymbol{\omega}$ one does not need a finite rotation θ, but an infinitesimal rotation $\delta\theta$. Now it so happens that two arbitrary infinitesimal rotations $\delta\theta_1$, and $\delta\theta_2$ do commute even though finite rotations do not. Since the definition of $\boldsymbol{\omega}$ involves only $\delta\theta$, $\boldsymbol{\omega}$ is a valid vector. In other words, we should not think of the angular velocity $\boldsymbol{\omega}$ as the time derivative of some vector angle θ, but rather as an infinitesimal rotation vector divided by an infinitesimal time interval.

Conservation is invariance in time

S62 *Let me ask you another thing that sometimes confuses me. Is a scalar quantity always conserved?*

T62 Be careful again about the words you use in physics. We call a quantity a scalar if it has no direction in space. To put it in the language of transformations that we used for defining a vector, we can say that a quantity is a scalar if it does not change under rotations of the co-ordinate axes. We say a quantity is conserved if it does not change in time, i.e. if its initial value is the same as its later (or earlier) value. The two words mean different things. Sometimes a common word 'invariant' may be used. In that case, you must try to see what it is supposed to mean from the context. A scalar is an 'invariant' under space rotations, a conserved quantity is an 'invariant' in time.

S63 *I get the point. Any examples so that I can absorb it better.*

T63 You can think of many examples yourself. Here are some. Take kinetic energy of a particle falling under gravity. This is a scalar but not a conserved quantity. Total energy of an isolated system is both a scalar and a conserved quantity. Similarly total charge of an isolated system is both a scalar and a conserved quantity. Thus a scalar may be conserved or may not be conserved. It depends on what the quantity is and what the situation is.

S64 *But is the opposite true? Is a conserved quantity necessarily a scalar?*

T64 Not at all. You know angular momentum is a vector. And if there are no external torques, this vector quantity is conserved, i.e. both the magnitude and direction of angular momentum do not change in time in the absence of external torques. So here we have an example of a conserved vector quantity.

Equivalence of rotated frames

S65 *Let me now ask something else. We are always cautioned that in any equation involving vectors, we should see that the left and right sides of the equation have the same character, i.e. if the left side is a vector (a scalar), the right side should also be a vector (a scalar). Why is this so?*

T65 Is it not obvious? Suppose I write a stupid thing like a scalar $b = \boldsymbol{A}$. The left side is just one number, whereas the right side refers to a triplet of numbers. The equality is obviously absurd. In the same way I cannot equate a tensor of rank 2 to a vector because the former has 9 components and a vector has only three. And so on.

S66 *No, but suppose I write an equation like $b = A_x$ where b is a scalar and A_x is the x-component of a vector. Now there is no problem of the kind you raised. Each side contains just one number. Or I may write, say, $T_{iy} = S_i$. Here T_{iy} is T_{ij}, with $j = y$, $T_{ij}(i, j = x, y, z)$ being a second rank tensor, and $S_i(i = x, y, z)$ is a vector. Here again the number of quantities on either side is the same (equal to 3). Are such equations possible in physics?*

T66 You must realize that it is not merely a question of counting how many quantities there are on each side. The equations you have suggested could appear in some particular frames chosen for a problem. But if you want the equations to be expressed in a form that looks the same in any frame, no matter how it is orientated, then these equations are not appropriate. (Equations having the same form in different frames† are sometimes described as 'form invariant' or 'covariant' equations.)

S67 *I did not get you. Can you illustrate what you mean by an equation having the same form in different rotated frames?*

T67 Well, an equation like $b = \boldsymbol{A} \cdot \widehat{\boldsymbol{n}}$ (where $\widehat{\boldsymbol{n}}$ is determined by some direction in the problem) has the same form in different frames. This is because both sides of the equation have the same character under rotation (both are scalars). The form invariance of the equation is also evident when we write the scalar product in the component form

$$b = A_x n_x + A_y n_y + A_z n_z = A'_x n'_x + A'_y n'_y + A'_z n'_z.$$

Now if we choose our co-ordinate axes such that $\widehat{\boldsymbol{n}}$ is along, say the x-axis then this equation becomes $b = A_x$. This equation is not wrong; it is only true in a particular frame. It does not exhibit the form invariance that the original equation does.

S68 *Your example of form invariance looks to be rather trivial. In the equation $b = \boldsymbol{A} \cdot \widehat{\boldsymbol{n}}$, each side is a scalar and does not change under rotations anyway. So it is bound to form invariant.*

T68 You are right. I shall give you a better example. Consider the equations

$$A_x = B_y C_z - B_z C_y,$$
$$A_y = B_z C_x - B_x C_z,$$
$$A_z = B_x C_y - B_y C_x.$$

The quantities occurring here are not scalars; they are components of vectors. The components of these vectors will differ from frame to frame. Yet I claim that these equations are form invariant.

S69 *How do you say that?*

T69 The way to see that is to make use of the transformation equations given in $T59$ for any general rotation. For simplicity you may restrict to rotations about the z-axis as in $T55$. You should be able to see that

† The phrase 'different frames' here, and up to $T78$, refers to frames which are rotated with respect to each other.

the equations in $T68$, when expressed in terms of primed.quantities using the transformation equations given in $T55$, lead to

$$A'_x = B'_y C'_z - B'_z C'_y,$$
$$A'_y = B'_z C'_x - B'_x C'_z,$$
$$A'_z = B'_x C'_y - B'_y C'_x.$$

Evidently, the form of these equations is identical to that of the equations in $T68$. I hope the meaning of 'form invariance' is now clear to you. To repeat, it is not that the quantities remain the same in the two frames (though that happens if the quantities are scalars); it is the form of the relations between the quantities which remain unchanged.

S70 *It occurs to me that I need not go into the explicit transformations in T55 to prove the relations in T69. I notice that* $(B_y C_z - B_z C_y)$ *is the x component of* $\boldsymbol{B} \times \boldsymbol{C}$, *i.e. the equations in T68 could be rewritten as*

$$A_x = (\boldsymbol{B} \times \boldsymbol{C})_x,$$
$$A_y = (\boldsymbol{B} \times \boldsymbol{C})_y,$$
$$A_z = (\boldsymbol{B} \times \boldsymbol{C})_z$$

In other words these equations are nothing but $\boldsymbol{A} = \boldsymbol{B} \times \boldsymbol{C}$. *I can now take the components of this vector equation in the* S' *frame and get*

$$A'_x = (\boldsymbol{B} \times \boldsymbol{C})'_x, A'_y = (\boldsymbol{B} \times \boldsymbol{C})'_y, \; A'_z = (\boldsymbol{B} \times \boldsymbol{C})'_z,$$

which lead to the relations in T69.

T70 Excellent! That is precisely the great advantage of writing equations in vector form. Vector equations are frame-independent. And so if you take their components in different frames, the resulting equations are bound to have the same form, i.e. equations written in vector form automatically guarantee form invariance.

S71 *Can you summarize this discussion on form invariance of equations?*

T71 In a form invariant equation, the transformation property (under rotation) of the two sides must match. Thus an equation like $C_i = A_i + B_i$ is form invariant because each side transforms similarly. Likewise an equation like $T_{ij} = S_{ij} + A_i B_j$ is form invariant because each side is a tensor of rank two. But an equation like $b = A_x$ or

$T_{iy} = S_i$ do not show form invariance. These equations could arise from form invariant equations in some particular frames.

Equations written in vector notation are independent of any frame and guarantee form invariance.

S72 *Can you give some examples to consolidate the point?*

T72 Remember the formula for the volume of a parallelepiped $V = \boldsymbol{A} \cdot \boldsymbol{B} \times \boldsymbol{C}$ in *T*27. Each side is a scalar. The formula makes no reference to any frame. If you write the formula in component form in two arbitrary frames S and S' it will obviously have the same form:

$$V = \begin{vmatrix} A_x & A_y & A_z \\ B_x & B_y & B_z \\ C_x & C_y & C_z \end{vmatrix} = \begin{vmatrix} A'_x & A'_y & A'_z \\ B'_x & B'_y & B'_z \\ C'_x & C'_y & C'_z \end{vmatrix}. \tag{1}$$

However, if one goes to a particular frame where $\boldsymbol{A}$ is along the x-axis, one gets

$$V = A_x(B_yC_z - B_zC_y). \tag{2}$$

Equation (2) is not wrong but is true only in a particular choice of frames. In an arbitrary frame V is not given by Eq. (2). Thus Eq. (2) does not have the form invariance that the original formula shows.

As another example consider the magnetic field of a point dipole. The magnetic field $\boldsymbol{B}$ at $\boldsymbol{r}$ due to a point dipole of the dipole moment $\boldsymbol{m}$ located at $\boldsymbol{r}'$ is given by

$$\boldsymbol{B}(\boldsymbol{r}) = \frac{3\widehat{\boldsymbol{n}}(\widehat{\boldsymbol{n}} \cdot \boldsymbol{m}) - \boldsymbol{m}}{|\boldsymbol{r} - \boldsymbol{r}'|^3}$$

where $\widehat{\boldsymbol{n}}$ is a unit vector in the direction pointing toward $\boldsymbol{r}$ from $\boldsymbol{r}'$. If, however, we choose our co-ordinate frame such that its origin is at the dipole and the z-axis points in the direction of $\boldsymbol{m}$ then the field components are given by

$$B_x(\boldsymbol{r}) = 3m\frac{xz}{r^5}, B_y(\boldsymbol{r}) = 3m\frac{yz}{r^5}, \; B_z(\boldsymbol{r}) = 3m\frac{z^2}{r^5} - \frac{m}{r^3}.$$

Again, this formula, although not wrong, does not have the form invariance that the formula in vector notation guarantees.

S73 *I now understand the notion of form invariance of equations in different frames. But perhaps I should have asked you this in the beginning; why do you demand this form invariance at all?*

T73 Because I believe that two frames S and S' having different orientations of axes must be completely equivalent as far as laws of physics or geometry are concerned. There is no preferred orientation of axes in space. So, I should always be able to express laws in a form which looks the same in different frames.

S74 *Is that just a matter of faith?*

T74 In a sense, yes. It is equivalent to saying that space is isotropic, i.e. there is no intrinsically preferred direction in space. I suppose experience bears this out. Things do not seem to behave differently in two houses just because they are oriented differently.†

S75 *This business of equivalence of rotated frames confuses me a little. Quantities do change from one frame to another. Yet you insist the frames are equivalent.*

T75 Of course, quantities may take different values in different frames. We have seen that in *T*59. Equivalence of frames means that the laws of physics make no distinction between frames of different orientation, i.e. the relations connecting various quantities do not change their form from one frame to another. In other words physical equivalence of different frames $\Leftrightarrow$ form invariance of mathematical equations which express the laws of physics.

S76 *In practice one does seem to have preferred frames. For example, if I am considering the motion of a charged particle in a constant magnetic field* ***B****, it makes good sense to align one of the axes of my frame along* ***B****. An arbitrary orientation of axes will be very inconvenient to work with.*

T76 Be careful. Do not equate 'a convenient frame' to 'an intrinsically preferred frame'. For a given problem a frame may be convenient for handling your equations. That does not make this frame intrinsically preferred over others from the point of view of basic laws of physics, and the basic characteristics and correlations of phenomena that these laws describe.

---

† This observation, strictly, implies translational and rotational invariance of the laws of physics.

S77 *But while working in a convenient frame S, I may not see its equivalence with some other frame S'. In fact I may not be interested in any other frame at all.*

T77 That may well be so. That is precisely what we saw in *T*72. When you are working out things in a particular frame, making convenient choices of this and that (without losing generality), plugging in numbers for quantities etc. this equivalence will not stand out. But that does not mean your laws do not respect the equivalence of different frames. It is just that it is not manifest in your procedure. If somebody insists, you must be able to work out your steps backwards and show that your basic equations are expressible in a form that shows this equivalence manifestly.

Reflection of axes

S78 *I have heard about pseudoscalars and pseudovectors. Can you explain what these terms mean?*

T78 This has to do with how quantities behave under the reflection of the axes, i.e. under the transformation

$$x' = -x,$$
$$y' = -y,$$
$$z' = -z.$$

You can easily see that under these reflections, a right-handed system is turned into a left-handed system and vice versa (see Fig. 2.10).

S79 *What about these transformations?*

T79 You see, under reflection, the components of a position vector $\boldsymbol{r}$ transform as given in *T*78. A vector $\boldsymbol{A}$ whose components transform similarly under reflection (i.e. $A_i' = -A_i, i = x, y, z.$) is called a polar vector. But sometimes one may have a vector $\boldsymbol{B}$ whose components transform exactly like those of $\boldsymbol{r}$ under rotations but behave differently under reflection, i.e. $B_i' = +B_i$, $i = x, y, z$. Such a vector is called an axial vector or a pseudovector.

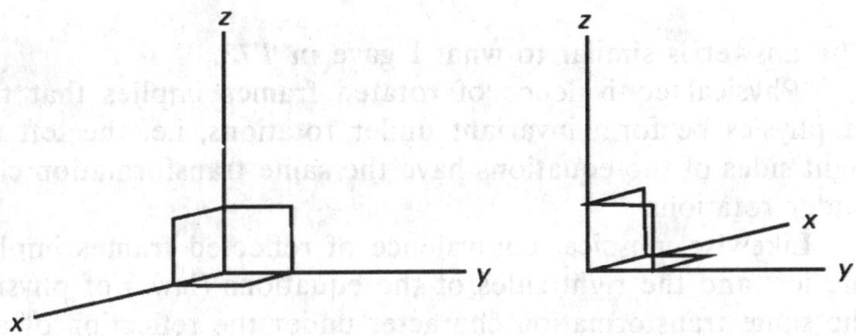

Fig. 2.10 (a) A right-handed co-ordinate system. (b) A left-handed co-ordinate system.

S80 *Do we come across such vectors in practice?*

T80 Yes. First take the linear momentum vector

$$\boldsymbol{p} = m\frac{d\boldsymbol{r}}{dt}.$$

$\boldsymbol{p}$ is clearly a polar vector. Now, consider the familiar angular momentum vector $\boldsymbol{L} = \boldsymbol{r} \times \boldsymbol{p}$. Under reflection components of both $\boldsymbol{r}$ and $\boldsymbol{p}$ reverse their signs. Clearly the components of $\boldsymbol{L}$ will not change under reflection. So $\boldsymbol{L}$ is an axial vector. A cross product of two polar vectors is an axial vector A cross product of an axial and a polar vector is a polar vector. And so on.

S81 *And what is a pseudoscalar?*

T81 Consider the dot product of a polar vector $\boldsymbol{P}$ with an axial vector $\boldsymbol{Q}$. Clearly $\boldsymbol{P} \cdot \boldsymbol{Q}$ is invariant under rotations like any other scalar. Under reflection, however, the components of $\boldsymbol{P}$ change sign but those of $\boldsymbol{Q}$ do not. Consequently, $\boldsymbol{P} \cdot \boldsymbol{Q}$ changes sign under reflection, unlike an ordinary scalar. Such a quantity is called a pseudoscalar.

You must have realized how easy it is to decide if something is a scalar or a pseudoscalar; or if it is a vector or a pseudovector. The transformation property under reflection is so simple.

S82 *Yes. But what is the big point? Why make so much fuss about how quantities behave under reflection of axes?*

T82 The answer is simple. We want the laws of physics to have the same form as we go from a right-handed to left-handed system. To ensure this, we must know how quantities transform under reflection.

S83 *Why do you make that requirement?*

T83 The answer is similar to what I gave in *T*73.

Physical equivalence of rotated frames implies that the laws of physics be form invariant under rotations, i.e. the left and the right sides of the equations have the same transformation character under rotations.

Likewise, physical equivalence of reflected frames implies that the left and the right sides of the equations (laws) of physics have the same transformation character under the reflection of axes.

S84 *But why do you believe in the physical equivalence of reflected frames?*

T84 Because we believe that nature has no preferred handedness, i.e. physical laws should not distinguish between left-handed and right-handed frames (for an exception, see T85).

S85 *Any examples?*

T85 The motion of a charged particle in an electromagnetic field described by $\boldsymbol{E}$ and $\boldsymbol{B}$ is given by combining Newton's law with Lorentz's force equation,

$$\frac{d\boldsymbol{p}}{dt} = q\boldsymbol{E} + q\boldsymbol{v} \times \boldsymbol{B}.$$

Now the left side is a polar vector. Also we know that $\boldsymbol{E}$ is a polar vector and $\boldsymbol{B}$ is an axial vector. (see Box 2.2). The cross product of the polar vector $\boldsymbol{v}$ and the axial vector $\boldsymbol{B}$ is a polar vector. This means both sides of the equation transform similarly under reflection, which in turn means that the equation respects the equivalence of left-handed and right-handed systems.

This is true of all equations in electromagnetism. In fact this equivalence is respected by all forces of nature but for an important exception. The exception is the weak interactions of elementary particles which, for example, are responsible for nuclear beta decay.

S86 *One last question for curiosity. I have heard of something called four-vectors. What are these 'animals'?*

T86 Before I tell you about them I want you to note the word transformation that has come up throughout this chapter. Transformation generally means a change from the original to something new. In the sense that we have been using this word here it means change of axes, or better, change of frame of reference and the resulting change in the values of physical quantities. So far in this chapter we have considered two types of transformations—rotations of axes and reflection of axes. But there are other interesting transformations that one can consider. For example, one can consider transformation of quantities between two frames in relative motion. We have defined scalars, pseudoscalars, vectors, pseudovectors etc. in terms of their behaviour under rotational and reflection transformations. Analogously we can define 'scalars', 'vectors' etc. in terms of their behaviour under these other transformations.

In special relativity the three space co-ordinates x, y, z and the time co-ordinate t of an event transform to new co-ordinates when we go from one inertial frame S to another inertial frame S' in uniform relative motion. So, in place of the three-component position vector (x, y, z), we now have a four-component object $(t; x, y, z)$ that transforms to $(t'; x', y', z')$ when we go from S to S'. These transformations are known as Lorentz transformations. A set of four quantities (A_O; A_1, A_2, A_3) which transforms like (t; x, y, z) is a four-vector under Lorentz transformations. You will learn more about them later.

Box 2.2

The electric field $\boldsymbol{E}$ at a point $\boldsymbol{r}$ is determined by the charge distribution described by charge density ρ,

$$\boldsymbol{E}(\boldsymbol{r}) = \frac{1}{4\pi\epsilon_0}\int \frac{\rho(\boldsymbol{r}')(\boldsymbol{r}-\boldsymbol{r}')}{|\boldsymbol{r}-\boldsymbol{r}'|^3} d^3r'$$

In this expression, ρ and $|\boldsymbol{r}-\boldsymbol{r}'|$ are scalars and $(\boldsymbol{r}-\boldsymbol{r}')$ is a polar vector. Clearly, therefore, $\boldsymbol{E}$ also is a polar vector.

The magnetic field $\boldsymbol{B}$ at $\boldsymbol{r}$ is determined by the current distribution, described by current density $\boldsymbol{j}$, as

$$\boldsymbol{B}(r) = \frac{\mu_0}{4\pi}\int \frac{\boldsymbol{j}(\boldsymbol{r}')\times(\boldsymbol{r}-\boldsymbol{r}')}{|\boldsymbol{r}-\boldsymbol{r}'|^3} d^3r'.$$

The current density is related to the charge density ρ and the velocity of the flow of charges $\boldsymbol{v}$ by the relation $\boldsymbol{j} = \rho\boldsymbol{v}$. Since ρ is a scalar and $\boldsymbol{v}$ is a polar vector, $\boldsymbol{j}$ is a polar vector. It is now clear that the expression for $\boldsymbol{B}$ involves the cross product of two polar vectors, namely $\boldsymbol{j}$ and $\boldsymbol{r}-\boldsymbol{r}'$, and hence $\boldsymbol{B}$ itself must be an axial vector.

S87 *Can you give me the gist of what you have just said?*

T87 The notion of scalars, vectors and tensors is always with reference to a given type of transformation. Strictly, a statement that something is a scalar or a vector is incomplete. The complete statement is that a given quantity is a scalar or a vector under such and such transformation. In elementary physics one rarely talks of any transformations other than rotations or reflections. Hence, this very important point is often missed.

S88 *I will remember that. Shall we now come to something simpler?*

3

Kinematics

I. One-dimensional Motion

Instantaneous velocity

S1 *We talk of instantaneous velocity of a particle. I am not sure if I understand clearly what that means.*

T1 In the course of motion the position of a particle changes. For one-dimensional motion the position of the particle relative to some point (origin) O can be described by a variable x which is a function of time, i.e. $x = x(t)$. The definitions of velocity and acceleration are based on this function.

S2 *I know that. Velocity is simply displacement $x_2 - x_1$ divided by the time interval $t_2 - t_1$. Acceleration is the change in velocity $v_2 - v_1$ divided by time interval $t_2 - t_1$.*

$$v = \frac{x_2 - x_1}{t_2 - t_1}; \qquad a = \frac{v_2 - v_1}{t_2 - t_1}.$$

T2 Can you improve on your answer?

S3 *I do not understand your question. These are the elementary definitions of velocity and acceleration that are learnt in school. For example, if a train covers a distance of 200 km in 4 hours along a straight track, we say its velocity is 50 km/h in the direction of motion. Or, if a car moving in the east to west direction picks up speed from 0 to 10 ms^{-1} in 5 s, its acceleration is 2 ms^{-2} in the east to west direction. What is wrong with these definitions?*

T3 Well, tell me, can velocity and acceleration in your definitions depend on time?

S4 *Bodies need not have uniform velocity or acceleration. These could, of course, change with time.*

T4 Then, to which time does v or a as you have defined in $S2$ refer to; t_1 or t_2 or some other time?

S5 *Well, let me see. It could be t_1. It may well be t_2, or $(t_1 + t_2)/2$. Frankly, I am confused.*

T5 The confusion arises because you are loosely using the words 'velocity' and 'acceleration'. In your examples, strictly speaking, you should use the words 'average velocity or average acceleration over the given interval of time'. In other words, the definitions you have given in $S2$ are correct definitions of average velocity and average acceleration over an interval t_1 to t_2:

$$v_{ave} = \frac{x_2 - x_1}{t_2 - t_1}; \qquad a_{ave} = \frac{v_2 - v_1}{t_2 - t_1}.$$

Of course, if v is constant over the interval t_1 to t_2 then v_{ave} is obviously the same as v. Similarly, if a is constant over the time interval then a_{ave} is the same as a.

S6 *Then what are the defining relations for v and a if they are varying in time?*

T6 In that case, you will have to define velocity at an instant, the so-called instantaneous velocity $v(t)$. Similarly you will need to define instantaneous acceleration $a(t)$.

S7 *This instantaneous thing confuses me. To talk of velocity we need to talk of some displacement and the time interval required for the displacement. When we talk of velocity at an instant, we are not referring to any time interval or displacement. What is the meaning of velocity if you have not covered any distance?*

T7 You must get this absolutely clear. The meaning of instantaneous velocity $v(t)$ can be understood only in a limiting sense, and hence requires calculus. You see, a finite displacement $x_2 - x_1$, covered in a finite time interval $t_2 - t_1$ gives v_{ave}.

$$v_{ave} = \frac{x_2 - x_1}{t_2 - t_1}.$$

Now let $t_2 \to t_1$ i.e. $t_2 - t_1 \to 0$. Clearly, x_2 will also tend to x_1, i.e. $x_2 \to x_1$, or $x_2 - x_1 \to 0$. But you must have learnt in calculus that the limit of a ratio that has both the numerator and the denominator going to zero can be a well-defined number. This number denotes the velocity at time t_1 (which is the same as the velocity at t_2 since $t_2 - t_1 \to 0$).

$$v(t_1) = \lim_{t_2 \to t_1} \frac{x_2 - x_1}{t_2 - t_1}$$

The right side of this equation is the derivative of $x(t)$ with respect to t at $t = t_1$, i.e.

$$v(t)|_{t=t_1} = \frac{dx}{dt}\bigg|_{t=t_1}$$

Now, this is true whatever t_1 we may take. So we can just as well write

$$v(t) = \frac{dx}{dt}.$$

By analogous reasoning, the instantaneous acceleration $a(t)$ is defined as

$$a(t) = \frac{dv}{dt} = \frac{d^2x}{dt^2}.$$

S8 *Sometimes we talk of a particle instantaneously or momentarily at rest during its motion. How can a particle ever be at rest if it is undergoing motion?*

Box 3.1

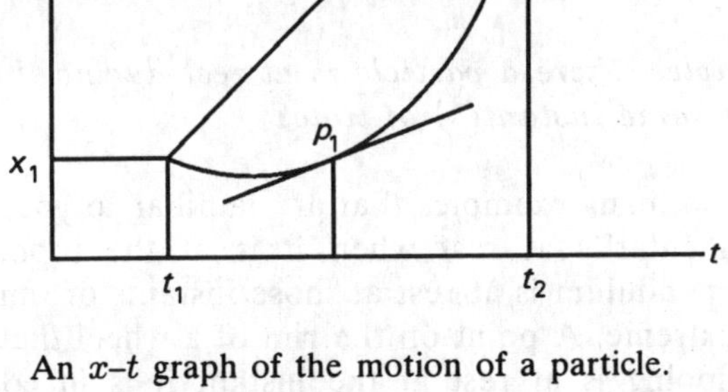

Fig. 3.1 An x–t graph of the motion of a particle.

The difference between average velocity and instantaneous velocity can be clearly understood if we look at the graph of a function $x(t)$.

Let the x-t graph for the motion of a certain particle be as shown in Fig. 3.1. The average velocity of the particle over the time interval $t_2 - t_1$ is $(x_2 - x_1)/(t_2 - t_1)$. This is just the slope of the chord P_1P_2.

On the other hand, the instantaneous velocity of the particle at a point, say P, is dx/dt, which is the slope of the tangent to the x–t graph at that point. This, clearly, is different from the slope of the chord in general.

T8 It simply means that a particle, during the course of its motion, has zero instantaneous velocity at some particular instant(s). This is quite possible, because the velocity of a particle in motion is, in general, a function of time, and may be zero at some instant(s). Put another way, the graph of function $x(t)$ may be such that at some points the tangent may have a zero slope. At such points, the instantaneous velocity of the particle will be zero. (See Fig. 3.2)

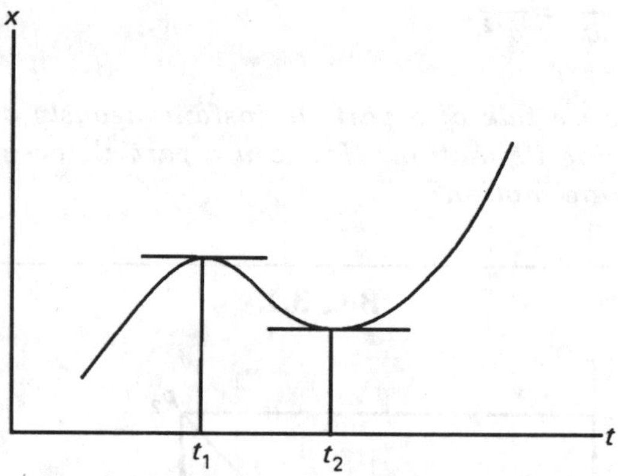

Fig. 3.2 Another x–t graph for the motion of a particle. The instantaneous velocity is zero at the instants t_1 and t_2.

S9 *Any examples where a particle is at rest during the course of its motion at some instant(s) of time?*

T9 There are so many examples that are familiar to you. A ball thrown up is momentarily at rest when it is at the topmost point. An oscillating pendulum is at rest at those instants of time when it is at its either extreme. A point on the rim of a wheel that rolls perfectly without slipping is at rest at the instant it is in contact with the ground.

S10 *When a particle is instantaneously at rest, i.e. its velocity is zero at some instant, is its acceleration also zero at that instant?*

T10 Not necessarily. It is a serious error to think so. Just go over the examples in T9. At the instant the ball is at the topmost point its velocity is zero but its acceleration continues to be g, the acceleration due to gravity. For an oscillating pendulum the velocity is zero at either end but its acceleration there is not zero; in fact the magnitude of acceleration is maximum at the end points. In the same manner the acceleration of a point on the rim of a rolling wheel, when the point is in contact with ground, is non-zero. (The last example is not of one-dimensional motion.)

S11 *But how can that be? I mean if $v = 0$, isn't $a = dv/dt$ also zero?*

T11 If v is zero, or has any constant value, at all times over an interval then certainly a is zero over the interval. But if $v = 0$ only at a particular instant, a may not be zero. If a function $v(t)$ is zero at $t = t_o$, it by no means implies that its derivative is zero at that point. This can be clearly seen from a v-t graph as shown in Fig. 3.3. The graph intersects the t-axis at $t = t_o$ so that $v(t_o) = 0$. The slope of the graph, $dv/dt = a$, at that point is, however, not zero.

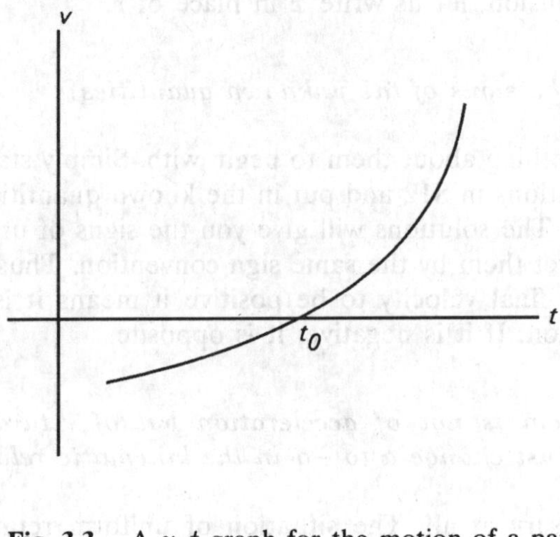

Fig. 3.3 A v–t graph for the motion of a particle.

Signs of kinematic quantities

S12 *Let us turn to the standard relations of kinematics in one-dimension,*

$$v = u + a\,t; \qquad s = ut + \frac{1}{2}a\,t^2; \qquad v^2 - u^2 = 2\,a\,s.$$

I want to be clear about the signs of the quantities here. Are the quantities u and v supposed to be positive?

T12 You must realize that all the quantities occurring in these equations are algebraic, i.e. they may be positive or negative (or zero).

S13 *Then how do we decide on the signs of these quantities in a given problem?*

T13 For a given problem choose any one direction to be the positive direction and stick to this choice throughout. Velocities and accelerations along the chosen direction are positive and those in the opposite direction are negative. In the above relations s stands for the displacement from the initial to the final position. It is positive if s is along the chosen direction and negative otherwise.

Sometimes s is confused with the magnitude of displacement. To avoid this confusion, let us write x in place of s.

S14 *But what about the signs of the unknown quantities?*

T14 You have to do nothing about them to begin with. Simply start with the kinematic relations in S12 and put in the known quantities with appropriate signs. The solutions will give you the signs of unknown quantities. Interpret them by the same sign convention. Thus, if the solution gives the final velocity to be positive it means it is along the chosen direction. If it is negative, it is opposite.

S15 *What if a problem is not of acceleration but of retardation? Surely, then, I must change a to $-a$ in the kinematic relations.*

T15 This is not necessary at all. The situation of uniform retardation is equally well included in the kinematic relations in S12. You do not need to change these expressions. All you must do is put the numerical value of a with the proper sign.

S16 *You mean I should take the value of a to be positive for speeding up and negative for slowing down. Right?*

T16 Be careful. Speeding up means acceleration is in the same direction as the instantaneous velocity. Slowing down means acceleration is in the opposite direction to the velocity.†

Now, often in simple problems where motion is unidirectional, one takes the positive direction to be the direction of velocity and then what you say is correct. But when in a problem the direction of motion changes, you have to be careful.

For example, consider a ball thrown up, which returns after some time. During the upward journey the ball slows down; its acceleration is in the opposite direction to that of velocity. During the downward journey it speeds up; its acceleration is in the same direction as that of its velocity.

Now if I choose the vertically upward direction to be the positive direction throughout the motion, the value of acceleration is negative throughout, although during downward journey the speed of the ball is increasing.

S17 *That is clear enough. Yet, when it comes to solving problems, I usually take the relations*

$$v = u + at; \qquad x = ut + \frac{1}{2}at^2; \qquad v^2 - u^2 = 2ax$$

for acceleration, and

$$v = u - at; \qquad x = ut - \frac{1}{2}at^2; \qquad v^2 - u^2 = -2ax$$

for retardation. That is the way it is given in some books too. Is there anything wrong?

T17 There is nothing wrong really. You are, in fact employing the same kinematic relations in either case. All you have done is to choose the direction of motion to be the positive direction and used the symbol a for the *underlinemagnitude* of a.

I repeat: the symbol a in S12 stands for acceleration which may have any sign; the symbol a in your relations (S17) stands for the magnitude of acceleration which cannot be negative. To avoid confusion you should use the symbol $|a|$ in your relations instead of a.

† This is true in one dimension. In three dimensions speeding up means acceleration vector has a positive component along the velocity vector.

S18 *All right. But for the actual solving of problems does it matter whether I use relations in S12 or those in S17?*

T18 If the problem involves unidirectional motion then there is little difference, but if the motion changes direction, it is better to use the relations in *S*12 because, then we do not have to worry about splitting a problem into one of acceleration and one of retardation.

S19 *Can you give me an example?*

T19 Consider a stone thrown upwards with an initial speed of 50 m s^{-1} from a cliff 375 m high from the ground. After what time will it hit the ground? (Take the acceleration due to gravity to be 10 m s^{-2}.)

S20 *It is easy. I shall first calculate the time the stone takes to go up, also calculate the height it reaches using the relations for retardation. Then I shall calculate the time it takes to come down to the ground using the relations for acceleration. I will add up the two times to get the answer:*

Upward motion $u = 50\,ms^{-1}$, $t = 5\,s.$

$v = 0\,ms^{-1}$ *Now use* $v^2 - u^2 = -2\,ax$

Use $v = u - at,$ *i.e.* $0 - 2500 = -20\,x,$

i.e. $0 = 50 - 10t,$ $x = 125\,m$

Downward motion $u = 0\,ms^{-1}$ *Use* $x = ut + \frac{1}{2}\,at^2,$

$x = 125 + 375 = 500\,m.$ *i.e.* $500 = 5\,t^2,$

$t = 10\,s.$

So the stone will hit the ground after 5 + 10 = 15 s.

T20 Your answer is correct. But the problem can be solved more efficiently using the relations in S12. We must choose some direction (up or down) to be the positive direction and stick to it throughout the problem. Let us choose upward direction to be positive. Then the given quantities are

$$u = +50\,m\,s^{-1}\ ,\ x = -375\,m,\ a = -10\,m\,s^{-2}$$

Use the relation $x = ut + (1/2)\,at^2,$

so that $-375 = 50t - 5t^2.$

This gives $t = +15s$ or $-5s$.

Since we are interested in finding when the stone hits the ground *after* it is thrown, we should take the positive solution. It agrees with your answer. I consider this method better because I do not have to worry about the upward and the downward parts of the motion separately. I got the answer from just one equation.

S21 *I agree with you, but I was a little surprised when you put $x = -375$ m in the above solution. x represents the height of the cliff. Should I not put it as equal to 375 m?*

T21 Be careful about the meaning of each quantity in the kinematic relations. Remember that x stands for the displacement from the initial to the final position of the object. This displacement is downward and we have taken the upward direction as the positive direction in the problem. So here x is a negative number.

S22 *I am now clear about the signs of x, u, v and a but what about the sign of t? I guess it must always be positive. Isn't that the reason you ignored the solution $t = -5s$?*

T22 No. t in kinematic equations can be either positive or negative.

S23 *But how can time be negative?*

T23 It is simply a question of which instant of time you choose to call zero. If I take this instant of time when you are reading this line as $t = 0$, all the events before this happened at negative times and all the events after this instant will occur at positive times. In other words, if the present is taken to be $t = 0$, all the past has $t < 0$ and all the future has $t > 0$.

S24 *I understand that time as such can be assigned negative values, but my question is whether the kinematic relations are true for $t < 0$.*

T24 The kinematic relations in S12 are true in the entire time domain during which the acceleration a is constant. The sign of t is immaterial.

S25 *Then why did you ignore the solution $t = -5s$?*

T25 In the kinematic relations of S12 note that at $t = 0$, $x = 0$ and $v = u$, i.e. $x(0) = 0$ and $v(0) = u$. Thus when we used these relations for our problem, the instant of time the stone is thrown up from the cliff corresponds to $t = 0$. The stone will hit the ground after that instant, so the time it hits the ground must be positive.

S26 *But what is the meaning of the solution $t = -5s$? Surely it should have some interpretation.*

T26 It has no interpretation for the given problem. We do not know what the situation was before the stone is thrown up, i.e. for $t < 0$. If the stone was at rest in the hands of the person who threw it at $t = 0$, we have $x = 0$ for $t < 0$. Thus for $t < 0$ the problem does not correspond to the one with uniform acceleration $= -10$ m s^{-2}. After it hits the ground, i.e., for $t > 15s$ and $x = -375$ m, again it is not a problem of uniform acceleration $= -10$ m s^{-2}. Hence the kinematic relations with $a = -10$ m s^{-2} are valid, in the given problem only for the time interval $0 \leq t \leq 15s$. Any solutions outside this interval must therefore be rejected.

S27 *Then what do we make of the solution $t = -5s$?*

T27 This solution will be perfectly applicable for a problem in which $a = -10$ m s^{-2} in the interval that includes $t = -5s$ and which has the same conditions at $t = 0$ as in the original problem.

S28 *I do not understand. Can you clarify?*

T28 Suppose I frame a problem like this: a stone thrown up from the ground passes the edge of a cliff 375 m high, with an upward speed of 50 m s^{-1}. Taking the instant it passes the cliff to be $t = 0$, what is the instant it was thrown up and at what instant will it return to the ground?

S29 *How do you solve this problem?*

T29 Exactly as in T20. The equation to solve it is identical to what we had before. But now both the solutions are interpretable. Both times correspond to $x = -375m$. $t = -5s$ is the instant the stone was thrown up from the ground and $t = 15s$ is the instant it returns to the ground.

S30 *Can you please consolidate everything that we have discussed so far?*

T30 (i) For problems on one-dimensional motion with uniform acceleration the kinematic relations are:

$$v = u + at; \qquad x = ut + \frac{1}{2}at^2; \qquad v^2 - u^2 = 2ax.$$

Note, in these equations u corresponds to velocity at $t = 0$ and the displacement x is zero at $t = 0$.

(ii) Do not change the algebraic form of these equations from situation to situation.

(iii) For a given problem, choose any convenient direction to be the positive direction and stick to the choice throughout the solution of the problem.

(iv) Having made this choice, substitute the known data for u, v, x, a, etc. In so doing do take care whether to put a positive number or a negative number for them. Put a positive value when the given quantity is along the positive direction that you have chosen a negative value if it is in the opposite direction.

(v) Solve for the unknowns. The quantities will come out with appropriate signs. Interpret the signs of the quantities that you obtain as in (iv), i.e. if the quantity turns out to be positive, it is along your chosen positive direction; if it turns out to be negative it is in the opposite direction.

(vi) Be reminded that the kinematic relations above are valid only for the constant a. If a is given to be a constant only over an interval t_1 to t_2, then the values of t outside this interval are unacceptable.

Origin in space and time is a matter of choice

S31 *I would like to know one thing. In a given problem which instant of time should I take to be* $t = 0$? *And which point should be taken as the origin, i.e.* $x = 0$?

T31 You will no doubt agree that what really matters is the elapsed time and the distance covered in that time. Which particular time you call $t = 0$ and which particular point in space you take as the origin is entirely a matter of choice.

Any instant of time during motion can be taken to be $t = 0$. $t = 0$ does not necessarily correspond to the beginning of the motion (See T28). It is customary to call u in the relation $v = u + at$ as 'initial' velocity. The word 'initial' here can be misleading. To avoid possible confusion always think of u above as v at $t = 0$.

For example, in the problem in *T*28 we took the edge of the cliff as $x = 0$ and the instant the stone passes the edge of the cliff as $t = 0$. But we could just as well take the instant when the stone is thrown from the ground as $t = 0$ and the ground as $x = 0$. Or else we could take the instant the stone is at the topmost point to be $t = 0$ and the topmost point as $x = 0$. Indeed we could take any point on the path as $x = 0$, and the instant the stone passes this point (in a specified direction) to be $t = 0$.

S32 *Let us see if we get the same answer if we make a different choice.*

T32 Suppose I choose the first alternative, i.e. $x = 0$ on the ground and $t = 0$ when the stone is thrown up. Then, when the stone passes the edge of the cliff it has $x = 375$m and $v = +50$ ms^{-1}. Use of the relation $v^2 - u^2 = 2ax$ gives u, the velocity with which the stone is thrown from the ground, to be 100 ms^{-1}. Let t' be the time the stone passes the edge of the cliff. The relation $v = u + at$ gives t' to be 5s. Next, let us use $u = 100$ ms^{-1} and $x = 0$ in the relation $x = ut + (1/2)at^2$. The resulting quadratic equation, $0 = 100t - 5t^2$, has two solutions: $t = 0$ corresponds to the initial instant of throw and $t = 20s$ to the instant the stone returns to the ground.

S33 *We seem to be getting different answers from what we obtained in T29.*

T33 The absolute values of the times of throw and return to the ground will, of course, not agree with our previous answers because of the shift of the zero of time. You must note, however, that the elapsed time between any two events comes out the same by either method. For example, the time between the throw and return to the ground is 20s in either case.

S34 *In all the discussion above we have taken $x = 0$ when $t = 0$. Is that necessary?*

T34 Yes, if we use the kinematic equations of the form given in T30. This condition, as you can see, is inbuilt in these relations. Of course, it is possible to modify these equations so that they can be used even when our reference point does not correspond to $t = 0$, $x = 0$.

S35 *How does one modify them?*

T35 In our kinematic equations t stands for the time elapsed from the instant $t = 0$ when the velocity of the object is u. Similarly x represents the displacement from the point $x = 0$. Let us choose

$t = t_0$ and $x = x_0$ as our reference, instead of $t = 0$ and $x = 0$; u, now, denotes the velocity at $t = t_0$ rather than at $t = 0$. For notational uniformity let us replace u by v_o so that the velocity at $t = t_0$ is denoted by v_0.

With the new choice, the elapsed time is $t - t_0$ and the displacement is $x - x_0$. Hence the kinematic equations modify to

$$v = v_0 + a(t - t_0),$$

$$(x - x_0) = v_0(t - t_0) + \frac{1}{2}a(t - t_0)^2$$

$$v^2 - v_0^2 = 2\,a(x - x_0).$$

You can explicitly check here that at $t = t_0$, $x = x_0$ and $v = v_0$, i.e. $x(t_0) = x_0$ and $v(t_0) = v_0$.

S36 *What, then, is the form of the one-dimensional kinematic relations that I should finally remember?*

T36 You may remember the relations in T35. It is best to view them as relations connecting the values of the two functions $x(t)$ and $v(t)$ at any time t to their values at $t = t_0$. What you must realize is that there is no special status to t_0 in these equations. By that I mean t and t_0 are any two instants of time. To highlight this point let us write t_0 as t_1 and t as t_2 and rewrite the relations above as

$$v(t_2) = v(t_1) + a(t_2 - t_1),$$

$$x(t_2) - x(t_1) = v(t_1)(t_2 - t_1) + \frac{1}{2}a(t_2 - t_1)^2,$$

$$v(t_2)^2 - v(t_1)^2 = 2\,a\,(x(t_2) - x(t_1)).$$

The relations are true for any t_1 and t_2 which belong to the interval of time during which the motion has a uniform acceleration a.

Variable acceleration

S37 *You have been emphasizing that kinematic relations are valid for constant acceleration. What happens in the case of time-varying acceleration? Can I simply put in a time dependence in a, i.e. use $a(t)$ in place of a in the same relations? I mean,* *are the following relations O.K. for the case of time-varying acceleration?*

$$\left.\begin{aligned} v(t) &= v_0 + a(t)(t - t_0), \\ x(t) &= x_0 + v_0(t - t_0) + \frac{1}{2}a(t)(t - t_0)^2, \\ v(t)^2 &= v_0^2 + 2a(t)(x(t) - x_0)^2 \end{aligned}\right\} ?$$

T37 No. If you know a general result or relation, you can, of course, always go to a special case of the general result. But the converse is not true. You cannot go from a special case to the general case in the *ad hoc* manner that you have adopted.

S38 *Well, then, how do I go about treating the general case?*

T38 That is an important question. Actually, what you must realize is that kinematic relations such as those in T35 follow from the definitions of velocity and acceleration. So to derive kinematic relations for the time-varying case simply go back to these definitions:

$$v = \frac{dx}{dt}; \qquad a = \frac{dv}{dt}.$$

The definition of a can also be put in an alternative form,

$$a = \frac{dv}{dt} = \frac{dv}{dx}\frac{dx}{dt} = v\frac{dv}{dx}.$$

S39 *Well, how do you get the kinematic relations from these definitions?*

T39 Before I answer that, note two things.

i) Definitions of v and a are differential equations; quantities on either side of these equations refer to the same time t.
ii) The kinematic relations (in *T*35), as already emphasized, relate the functions $x(t)$ and $v(t)$ to their values at a particular time $t = t_0$,

Obtaining (ii) from (i) is what one means by integrating differential equations, or solving them with given initial (or boundary) conditions.

S40 *O.K. then, how does one obtain (ii) from (i)?*

T40 Consider first the simplest situation. A body is at rest at a point

x_0 at some instant t_0, and its acceleration is zero throughout some time interval t_1 to t_2 that includes the time instant t_0.

S41 *But that is not the situation we are talking about. We are talking about a body moving with varying $v(t)$ and perhaps with varying $a(t)$.*

T41 Remember what I said in *T*37. A simple situation that is a special case of a general situation is always included in the latter. So you should be able to handle the simple situation with the general method.

S42 *All right. Please proceed.*

T42 Let us see what is given. We know that $a = 0$ throughout the time-interval t_1 to t_2. Further we have what are called initial conditions, $x(t_0) = x_0$ and $v(t_0) = 0$.

S43 *And what do we have to find out?*

T43 We must find out x and v for all t in the given interval. This is easily done by integrating the defining relations.

Start with $dv/dt = a$. Now $a = 0$, therefore $dv/dt = 0$ for all t. Clearly, $v(t)$ must be a constant, $v(t) = k$.

It is also given that $v(t_0) = 0$. The constant k must, therefore, be zero. Hence $v(t) = 0$.

Now, use the definition of velocity $dx/dt = v$. Since $v(t) = 0$ as already proved, we have $dx/dt = 0$. This gives $x(t)$ = constant. But it is given that $x(t_0) = x_0$. Therefore $x(t) = x_0$.

We have thus solved the problem obtaining $v(t) = 0$ and $x(t) = x_0$.

S44 *I do not see what you have got out after all this. I knew already that the body would be at rest at x_0 all the time.*

T44 You are right. I went through this exercise just to show you that solving the differential equations in T38 for this simple case leads to the expected results.

Let us now consider another simple situation. A body is at position x_0 and has a velocity v_0 at time t_0. The acceleration of the body is zero throughout the interval t_1 to t_2, which includes t_0.

S45 *Well, let me try to imitate what you did in T42 and T43. First* *we are given that $a = 0$ throughout the time interval. Further, the initial conditions are $x(t_0) = x_0$ and $v(t_0) = v_0$.*

T45 That is correct. What do you do next?

S46 *From $dv/dt = a$ we get $dv/dt = 0$, which gives $v(t) =$ constant. Using the information $v(t_0) = v_0$ we can conclude that this constant must be v_0. Hence $v(t) = v_0$.*

T46 Very good. Go on.

S47 *Next I use the relation $dx/dt = v$. With v that we have obtained, this relation becomes $dx/dt = v_0$. Integration of this equation gives $x(t) = v_0t + C$, where C is a constant. Now I shall use the other piece of information, namely $x(t_0) = x_0$. Substitution gives $x_0 = v_0t_0 + C$, which gives $C = x_0 - v_0t_0$. This, finally, gives me the result $x = v_0t + x_0 - v_0t_0$.*

T47 Excellent. Let me just rewrite your results: $v(t) = v_0$ and $(x - x_0) = v_0(t - t_0)$. These are the relations that you anyway know for uniform motion. So, the method gives expected results in this simple situation also.

Let us, now, consider the motion of a particle with uniform acceleration. In this case $a(t) = a$, where a is a constant. The initial conditions can be taken as $x(t_0) = x_0$ and $v(t_0) = v_0$.

Now $dv/dt = a$. Since a is a constant, the integration of this equation gives $v(t) = at + \alpha$, where α is a constant.

The use of the initial condition $v(t_0) = v_0$ gives α to be $v_0 - at_0$. Therefore,

$$v(t) = (v_0 - at_0) + at = v_0 + a(t - t_0).$$

Next, $dx/dt = v(t) = v_0 + a(t - t_0)$.

Integrating both sides we get

$$x(t) = (v_0 - at_0)t + \frac{1}{2}at^2 + \beta, \text{ where } \beta \text{ is a constant.}$$

At $t = t_0$ this relation gives

$$x_0 = (v_0 - at_0)t_0 + \frac{1}{2}at_0^2 + \beta,$$

which in turn determines β. When β, thus obtained, is substituted back in the expression for $x(t)$ we obtain

$$x(t) = x_0 + v_0(t - t_0) + \frac{1}{2}a(t - t_0)^2.$$

So, we have got the exact kinematic relations for constant acceleration that you are familiar with (See T35).

S48 *This seems to be a very clear way of deriving kinematic relations. I now appreciate that the kinematic relations are nothing but the solutions of the defining equations for v and a with given initial conditions. But many books give a much simpler derivation of these relations. The relation $v(t) = v_0 + a(t - t_0)$ is obtained straight from the definition of acceleration*

$$a = \frac{v - v_0}{t - t_0}.$$

To obtain the relation for displacement one starts with the definition of average velocity

$$v_{ave} = \frac{x - x_0}{t - t_0}.$$

Next we take

$$v_{ave} = \frac{v_0 + v}{2} = v_0 + \frac{1}{2}a(t - t_0).$$

to obtain

$$(x - x_0) = v_0(t - t_0) + \frac{1}{2}a(t - t_0)^2$$

Is this derivation correct?

T48 This is a perfectly correct procedure provided that a is constant.

S49 *Just where has the fact that a is constant gone into in the above derivation?*

T49 At two places.

First, I have repeatedly stressed that $(v - v_0)/(t - t_0)$ is the definition of average acceleration, not instantaneous acceleration. But, of course, if a is constant it is the same thing as a_{ave}.

Second, you must realize that the formula $v_{ave} = (v_0 + v)/2$ that you have employed above is true if a is constant, but not for general motion.

S50 *That surprises me. Isn't the average of two numbers always calculated by adding them and dividing the sum by two?*

T50 Here we are not talking about the average of two numbers; we are talking about the average of a quantity over an interval. Simple-minded relations like $v_{ave} = (v_0 + v(t))/2$ and $a_{ave} = (a(t_0) + a(t))/2$ are not true for arbitrary motion.

S51 *This seems to be an important point. Please elaborate on it.*

T51 I shall repeat what I have just said. The relations

$$v_{ave} \equiv \frac{x - x_0}{t - t_0}; \; a_{ave} \equiv \frac{v - v_0}{t - t_0}$$

are always true; in fact they are definitions of v_{ave} and a_{ave} as I have indicated by the three lines. But the relations

$$v_{ave} = \frac{v_0 + v(t)}{2} \; ; \; a_{ave} = \frac{a_0 + a(t)}{2}$$

are not true in general.

S52 *Then how does one calculate these averages in general?*

T52 For v_{ave}, begin with the definition $dx/dt = v(t)$. The integration of this relation gives

$$x(t) = \int_{t_0}^{t} v(t')dt' + constant.$$

Now, at $t = t_0$, $x = x_0$ which gives the constant equal to x_0.

$$\therefore \; x(t) - x_0 = \int_{t_0}^{t} v(t')dt'.$$

This gives

$$v_{ave} = \frac{1}{t - t_0} \int_{t_0}^{t} v(t')dt'.$$

In the same way

$$a_{ave} = \frac{1}{t - t_0} \int_{t_0}^{t} a(t')dt'.$$

[In fact average of any quantity $f(x)$ over an interval x_1 to x_2 can be obtained as

$$f_{ave} = \frac{1}{x_2 - x_1} \int_{x_1}^{x_2} f(x)dx.]$$

S53 *Does the formula for v_{ave} given in T52 yield $(v_0 + v(t))/2$ for the case of constant a?*

T53 Yes, it does as is shown below.

For the constant a we have $v(t) = v_0 + a(t - t_0)$. Substituting this in the formula for v_{ave} given in T52, a straightforward integration gives

$$v_{ave} = v_0 + (1/2)a(t - t_0) = (v_0 + v(t))/2.$$

S54 *That is clear. Let us now return to my original question. What are the kinematic relations in the case of general time-dependent acceleration $a(t)$?*

T54 Here, then, comes the anti-climax. In the general case I cannot go far beyond the defining relations $v = dx/dt$ and $a = dv/dt$ themselves. I can, of course, write from T52

$$x(t) = \int_{t_0}^{t} v(t')dt' + x_0,$$

$$v(t) = \int_{t_0}^{t} a(t')dt' + v_0.$$

but these are nothing like the simple algebraic relations that you have in the case of constant acceleration.

You must appreciate though that, for any given $a(t)$, the problem of findingout the kinematic relations has been reduced to doing the integrals appearing above. Thus, although we do not have kinematic relations for an arbitrary $a(t)$, we have a systematic procedure to findout these relations for any given $a(t)$. Of course $a(t)$ has to be simple enough so that the integration is possible.

S55 *There is one thing that I noticed when you were trying to solve the differential equations $dx/dt = v$ and $dv/dt = a$. You always had <u>two</u> initial conditions $x(t_0) = x_0$ and $v(t_0) = v_0$. What is the reason behind this?*

T55 What you must realize is that acceleration alone does not determine x as a function of time completely. That is, one can have different types of motion under the same acceleration.

For example, motion under gravity is characterized by a uniform downward acceleration. But, of course, the initial position and initial velocity of a particle can be anything. I might throw a ball upwards from the ground or from a cliff. Or, I might just drop a stone from a certain height or shoot it down with some initial velocity. In short, if you simply specify the acceleration, it is not enough to determine what $x(t)$ and $v(t)$ are. One must have <u>two</u> additional pieces of information, the so-called initial conditions, such as the value of $x(t)$ and $v(t)$ at some instant $t = t_0$.

S56 *I understand that from the examples you give. But how are you sure that, in general after specifying a, two initial conditions are enough?*

T56 That follows from the mathematics of differential equations. The equation $d^2x/dt^2 = a$ is a linear second-order differential equation for $x(t)$. Mathematics tells us that for such an equation two initial conditions, such as $x(t_0)$ and $(dx/dt)_{t=t_0}$, are necessary and sufficient to determine $x(t)$.

Another way of looking at it is that we have two linear first-order differential equations $dx/dt = v$ and $dv/dt = a$. The second equation needs one initial condition (the value of v at $t = t_0$) to determine $v(t)$ completely. Having obtained $v(t)$, the first equation needs one more initial condition (the value of x at $t = t_0$) to solve for $x(t)$. Either way the answer is the same. We need two initial conditions to get $x(t)$ if a is specified. (Note, there is nothing wrong in specifying x at one instant and v at another instant.)

S57 *There is another thing I wanted to ask when you were clarifying the difference between average velocity and instantaneous velocity. Does a body with varying v ever possess velocity equal to v_{ave} during the course of its motion ?*

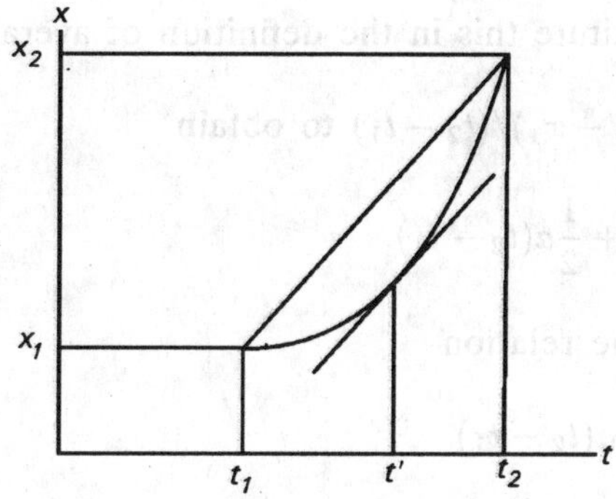

Fig. 3.4 An x–t graph for the motion of a particle.

T57 What you are asking is whether v_{ave} can equal $v(t)$ at some instant of time within the interval? The answer to this question is: yes, this is always true. You can understand this if you plot the x-t graph of motion. Suppose the graph for a certain motion is as shown in Fig. 3.4. Over the interval t_1 to t_2, $v_{ave} = (x_2 - x_1)/(t_2 - t_1)$, which is the slope of the chord connecting the end points of the interval. The instantaneous velocity of the particle at time t is dx/dt, i.e. the slope of the tangent of the x-t graph at that instant. Intuitively it is clear that for any differentiable curve (that is, a curve without breaks and kinks) you will always find a point on the curve where the tangent is parallel to the chord. This is proved rigorously in any calculus text and the result goes by the name of Mean Value Theorem. Hence there will always be some point $t = t'$ within the interval where

$$v(t') = \left.\frac{dx}{dt}\right|_{t=t'} = \frac{x_2 - x_1}{t_2 - t_1} \equiv v_{ave}.$$

Likewise, provided that v–t graph of motion is differentiable, there will always be some point t' within the interval such that

$$a(t') = \left.\frac{dv}{dt}\right|_{t=t'} = \frac{v_2 - v_1}{t_2 - t_1} \equiv a_{ave}.$$

S58 *Will the point t' be at the middle of the interval t_1 to t_2 necessarily?*

T58 No, not necessarily. It might happen to be so in some cases. For example, consider the case of uniform acceleration. In this case x–t graph is a parabola and the distance $x_2 - x_1$ travelled in time interval $t_2 - t_1$ is given by the relation

$$(x_2 - x_1) = v_1(t_2 - t_1) + \frac{1}{2}a(t_2 - t_1)^2.$$

We can substitute this in the definition of average velocity

$v_{ave} = (x_2 - x_1)/(t_2 - t_1)$ to obtain

$$v_{ave} = v_1 + \frac{1}{2}a(t_2 - t_1).$$

Now using the relation

$$v_2 = v_1 + a(t_2 - t_1)$$

we get $\quad v_{ave} = (v_1 + v_2)/2.$

On the other hand,

$$v(t)|_{t=\frac{t_1+t_2}{2}} = v_1 + a\left(\frac{t_1 + t_2}{2} - t_1\right)$$
$$= \frac{v_1 + v_2}{2}.$$

Thus, in the case of uniform acceleration, it is indeed true that v_{ave} over any interval t_1 to t_2 is equal to the instantaneous velocity at $t = (t_1 + t_2)/2$. But, of course, this will not be true in general.

x–t graphs

S59 *By the way, we are talking of linear motion and yet you are considering all kinds of curves for the x–t graphs. How?*

T59 You must confuse the shape of the x–t graph with the actual path of a particle in space. This is such an obvious point that it hardly needs any explanation. Yet so many times, quite unawares, students take x–t graphs for the shape of the path in space. For one-dimensional uniformly accelerated motion the x–t graph is parabolic. That by no means implies that the particle is moving along a parabolic curve; the motion is one dimensional, i.e. along a straight line.

Since you have made this kind of error, let me ask you a question. Is there anything wrong with the graph in Fig. 3.5?

S60 *Well, nothing seems to be wrong. The particle is moving in a circle in an anti-clockwise direction.*

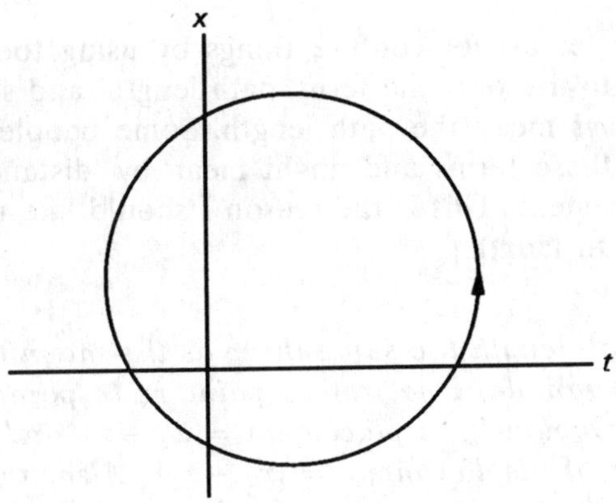

Fig. 3.5 An alleged x–t graph.

T60 So you have instantly forgotten what I just said.

S61 *Oh, it is an x–t graph and not the path of a particle in space. Well, let me see. The graph is wrong because t is shown to be negative.*

T61 There is absolutely nothing wrong with that. I have explained earlier that t can be positive, negative or zero.

S62 *I now see what is wrong with this graph. For a given t, $x(t)$ is not unique. A particle cannot possibly be at two different locations at the same time. Hence the given graph is absurd.*

T62 Excellent. But what is that arrow doing on the graph?

S63 *I do not know.*

T63 When you illustrate the trajectory of a particle in space, an arrow does make sense. It shows the direction of traversal of the path with increasing time. On an x–t graph it obviously makes no sense. So never put an arrow on an x–t graph.

Displacement and path length

S64 *Let us turn to something else. We use words like 'distance', 'path length', 'displacement'. Is the path length or the distance travelled the same as the magnitude of displacement?*

T64 No, not always. But let us not confuse things by using too many words. Let us agree to use only the term 'path length' and say that by 'distance' we always mean the path length. Some people might differ in the use of these terms and might mean by 'distance' the magnitude of displacement. This is the reason I should like to stick to only one term: path length.

S65 *Yes, but isn't the path length the same thing as the magnitude of displacement? For example, I go from a point x_1 to point x_2 as shown in Fig. 3.6. Obviously displacement $= x_2 - x_1$ and therefore, the magnitude of displacement $= |x_2 - x_1|$. Also, the path length $= |x_2 - x_1|$. The two are clearly equal.*

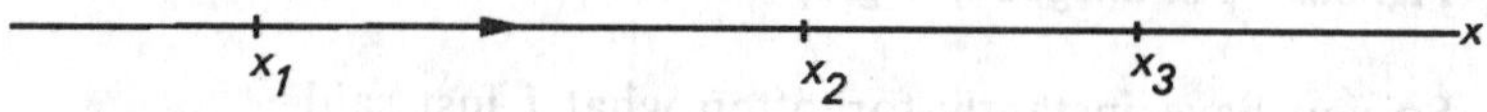

Fig. 3.6 A one-dimensional path of a particle.

T65 If during the motion you have not changed direction, then what you say is true. But what if I do not go from x_1 to x_2 directly; instead I go from x_1 to x_3 and then turn back to come to x_2? In this case the displacement from x_1 to x_2 is still $(x_2 - x_1)$ and the magnitude of displacement is still $|x_2 - x_1|$. But the path length, i.e. the length of the actual path traversed is clearly $|x_3 - x_1| + |x_3 - x_2|$. Obviously this is greater than the magnitude of the displacement.

S66 *From the way you have explained, it seems as if displacement has to do only with the positions of the particle at the beginning and at the end of the time interval of interest. It has nothing to do with the actual path taken by the particle in order to complete the journey. Is that right?*

T66 Perfectly. The path length, of course, depends on the actual path traversed by the particle. It is equal to the magnitude of the displacement if you have followed a straight path without any change in direction, not otherwise. As an extreme example I may start from x_1 and return to the same point following any kind of path I choose. The magnitude of displacement, then, is zero, but of course the path length is not zero. (I can experience that last statement in the aching of my legs!)

Average speed and average velocity

S67 *What can we say about average speed and the magnitude of average velocity?*

T67 It is really the same point.

$$\text{Average velocity} = \frac{x_2 - x_1}{t_2 - t_1}$$

$$\therefore \quad |\text{ Average velocity }| = \frac{|x_2 - x_1|}{t_2 - t_1}$$

$$= \frac{\text{magnitude of displacement}}{t_2 - t_1}.$$

On the other hand

$$\text{Average speed} = \frac{\text{path length}}{t_2 - t_1}..$$

Now, the path length $\geq$ magnitude of displacement.

$\therefore$ average speed $\geq$ magnitude of average velocity.

The equal sign holds only when, during motion on a straight line, you have not changed direction.

S68 *What about instantaneous speed and the magnitude of instantaneous velocity?*

T68 They are always equal.

S69 *Why?*

T69 Well, instantaneous quantities are defined in terms of vanishingly small intervals of time and vanishingly small displacements. During such small intervals, the length of the path covered is the same as the magnitude of the displacement. (The length of a vanishingly small arc is the same as that of the corresponding chord.) Consequently, the instantaneous speed of a particle is the same as the magnitude of instantaneous velocity.

S70 *Let me put the result in T67 symbolically. If I denote $|\delta x|$ by δs,* *then $v = dx/dt$ and $|v| = ds/dt$. The displacement in an interval t_1 to t_2 can be obtained as*

$$\int_{t_1}^{t_2} dx = \int_{t_1}^{t_2} v dt.$$

whereas the path length covered in the same interval is given by

$$\int_{t_1}^{t_2} ds = \int_{t_1}^{t_2} |v| dt.$$

Since the path length $\geq$ magnitude of displacement,

$$\int_{t_1}^{t_2} |v| dt \geq \left| \int_{t_1}^{t_2} v dt \right|.$$

Dividing by $t_2 - t_1$, I obtain

$$|v|_{ave} \geq |v_{ave}|.$$

Isn't that strange?

T70 Well, it just means that the order of the two operations involved, the operation of taking average over an interval and the operation of taking magnitude of a quantity, cannot be reversed. When this happens one says that the two operations do not commute.

S71 *We have been talking of one-dimensional motion. But the general motion of a body in physical space must be three-dimensional. What, then, is the relevance of studying motion in one dimension?*

T71 You are right. Bodies do move in three-dimensional space. But in some situations the motion along one direction may be important or relevant, motion along other two directions being uninteresting or unimportant. For example, motion of a train along a straight track is strictly a three-dimensional motion. But one may not be interested in how the train wobbles slightly along the breadth of the track or how it oscillates up and down on its suspension. We may simply be interested in its motion along the straight track for which we can employ the relations for one-dimensional motion.

II. Two- and Three-dimensional Motion

Vector kinematic relations

S72 *Let us turn to kinematics in two and three dimensions. In what important ways is it different from one-dimensional kinematics?*

T72 The key additional complication is due to the directionality of quantities like displacement, velocity, acceleration etc.

S73 *But these quantities too have direction in the one-dimensional case too.*

T73 True, but in one dimension there are only two directions possible. So we can take care of the direction of a quantity by the use of positive and negative signs. This will not work for two or three dimensions because here, obviously, the number of possible directions is infinite.

S74 *All right. What are the kinematic relations for constant acceleration in the three-dimensional case?*

T74 The generalization is straightforward. The one-dimensional relations that you are familiar with are now true for each cartesian component of position, velocity and acceleration.

$$v_x(t) = v_{0x} + a_x(t - t_0)$$

$$x(t) = x_0 + v_{0x}(t - t_0) + \frac{1}{2}a_x(t - t_0)^2$$

$$v_x(t)^2 = v_{0x}^2 + 2a_x(x(t) - x_0).$$

Similar relations hold for the y and the z components.

S75 *I suppose I can write these relations more compactly in vector notation as*

$$\boldsymbol{v}(t) = \boldsymbol{v}_0 + \boldsymbol{a}(t - t_0)$$

$$\boldsymbol{r}(t) = \boldsymbol{r}_0 + \boldsymbol{v}(t - t_0) + \frac{1}{2}\boldsymbol{a}\ (t - t_0)^2$$

$$\boldsymbol{v}(t)^2 = \boldsymbol{v}_0^2 + 2\ \boldsymbol{a}\cdot(\boldsymbol{r}(t) - \boldsymbol{r}_0).$$

T75 That is right. That in fact is the best way of expressing the kinematic relations in two or three dimensions. This is so since these relations relate position, velocity and acceleration vectors per se, without any reference to a co-ordinate system.

S76 *But how do you arrive at these vector kinematic relations?*

T76 That is no mystery. Remember what I said for the one-dimensional case. All kinematic relations follow from the definitions of velocity, acceleration etc. So you must be clear about the definitions of velocity and acceleration vectors in terms of the position vector.

S77 *How do you define them?*

T77 The velocity vector is simply the rates of change of the position vector. If $\boldsymbol{r}(t)$ is the position vector at time t and $\boldsymbol{r}(t+\delta t)$ is the position vector at $t+\delta t$, then the instantaneous velocity $\boldsymbol{v}(t)$ is given by

$$\boldsymbol{v}(t) = \underset{\delta t \to 0}{Lim} \frac{\boldsymbol{r}(t+\delta t) - \boldsymbol{r}(t)}{\delta t} = \frac{d\boldsymbol{r}}{dt}.$$

Likewise, the instantaneous acceleration vector is defined as the rate of change of the velocity vector

$$\boldsymbol{a}(t) = \underset{\delta t \to 0}{Lim} \frac{\boldsymbol{v}(t+\delta t) - \boldsymbol{v}(t)}{\delta t} = \frac{d\boldsymbol{v}}{dt}.$$

S78 *I am confused about the meaning of velocity vector at a certain time t. Afterall to define velocity there must be some displacement in space over a certain time interval. Similarly to define acceleration I must see how velocity changes over some time interval and divide the change by the time interval.*

T78 You are back to the same confusion that you showed earlier. Please address this barrier regarding instantaneous velocity and acceleration and try to overcome it.

Whatever you have said above refers to average velocity and average acceleration over a time interval.

$$\boldsymbol{v}_{ave} = \frac{\boldsymbol{r}(t_2) - \boldsymbol{r}(t_1)}{t_2 - t_1}; \qquad \boldsymbol{a}_{ave} \frac{\boldsymbol{v}(t_2) - \boldsymbol{v}(t_1)}{t_2 - t_1}.$$

Instantaneous velocity and instantaneous acceleration at some time

$t = t_1$ are the limits of these expressions as the interval $t_2 - t_1$ tends to zero. That leads us to the definitions given in T77. In short, if you are clear about the notions of limits and derivatives in calculus then $\boldsymbol{r}(t)$, $\boldsymbol{v}(t)$ and $\boldsymbol{a}(t)$ are simply different functions of time related by the equations in T77.

S79 *How does one get kinematic relations from these definitions?*

T79 In the one-dimensional case, we can integrate these equations for constant acceleration with suitable initial conditions.
Start with the equation

$$\frac{d\boldsymbol{v}}{dt} = \boldsymbol{a}.$$

For constant $\boldsymbol{a}$, the solution of this equation is

$$\boldsymbol{v}(t) = \boldsymbol{a}t + \boldsymbol{\eta}$$

Using the initial condition, $\boldsymbol{v}(t_0) = \boldsymbol{v}_0$, the constant of integration $\boldsymbol{\eta}$ can be evaluated. This gives

$$\boldsymbol{v}(t) = \boldsymbol{v}_0 + \boldsymbol{a}(t - t_0).$$

Next, integrating this equation and using the initial condition $\boldsymbol{r}(t_0) = \boldsymbol{r}_0$, we get

$$\boldsymbol{r}(t) = \boldsymbol{r}_0 + v_0(t - t_0) + \frac{1}{2}\boldsymbol{a}(t - t_0)^2.$$

Eliminating t between the two equations we can obtain a third relation

$$v(t)^2 = v_0^2 + 2\boldsymbol{a} \cdot (\boldsymbol{r}(t) - \boldsymbol{r}_0).$$

S80 *For the one-dimensional case we needed two numbers to specify the initial conditions. How many do we require here?*

T80 Although $\boldsymbol{r}_0$ and $\boldsymbol{v}_0$ appear to be two quantities, remember that they are vectors. They in fact, stand for six numbers. That is as it should be because the two vector equations in T77, are, in fact, six first-order differential equations. We know a first order equation requires for its solution the specification of one initial condition. So we need six initial conditions in all.

S81 *You have said that the kinematic relations are true for constant acceleration. In uniform circular motion, we have constant acceleration equal to v^2/r. Does that mean I can apply the relations in S75 to this case?*

T81 Be careful. The kinematic relations are true for the constant acceleration vector $\boldsymbol{a}$. This means both the magnitude and the direction of $\boldsymbol{a}$ should be constant. In uniform circular motion the magnitude of $\boldsymbol{a}$ is constant but not its direction. The direction of $\boldsymbol{a}$ is along the radius towards the centre and it is continuously changing during motion. The kinematic relations in S75 are, therefore, not applicable to uniform circular motion.

S82 *Does the direction of the acceleration vectors in uniform circular motion depend on the sense of rotation namely, clockwise or anticlockwise?*

T82 Obviously not. What appears clockwise to you will appear anticlockwise to somebody facing you. The inward direction of acceleration vector will be common to both.

S83 *Is the acceleration vector radially inward even when motion along a circle is not uniform?*

T83 No. The acceleration now has two components, one along the tangential direction and the other along the radially inward direction. The tangential component describes how the angular velocity of the object changes. The radially inward (centripetal) component of acceleration is, however, given by the same formula v^2/r where v is the instantaneous speed.

S84 *Are the vector kinematic relations in S75 true for uniform acceleration as well as uniform retardation? Or must I replace **a** by −**a** when the motion is retarded?*

T84 You must realize that $\boldsymbol{a} = d^2\boldsymbol{r}/dt^2$ is <u>the</u> definition of acceleration. It is true no matter whether the components of $\boldsymbol{a}$ along some axes are positive or negative. Replacing $\boldsymbol{a}$ by $-\boldsymbol{a}$ in this definition for the case of retarded motion is absurd. In the same way, the kinematic relations in *S*75, which are obtained from this definition, are true as long as all the components a_x, a_y, a_z are constant no matter what their signs. You should not change the algebraic form of these relations but only substitute positive or negative numbers for the components of $\boldsymbol{a}$ depending on what the situation is.

S85 *For non-uniform acceleration is it correct to replace the constant* $\boldsymbol{a}$ *by* $\boldsymbol{a}(t)$ *in the kinematic relations?*

T85 No. The kinematic relations are obtained by solving the defining relation $\boldsymbol{a} = d^2\boldsymbol{r}/dt^2$. Now, this relation is correct for arbitrary $\boldsymbol{a}$ but it is only when $\boldsymbol{a}$ is constant that the solution of this equation with initial conditions gives you the kinematic relations in *S*75. For arbitrary $\boldsymbol{a}(t)$, integration of the defining relation, if possible at all, will not give the relations you have suggested. Thus, your *ad hoc* procedure is wrong.

Acceleration and trajectory

S86 *For motion under constant acceleration* $\boldsymbol{a}$, *is it correct to say that the trajectory of a particle is always a straight line along the direction of* $\boldsymbol{a}$?

T86 Not true. All that you can say is that the component of velocity along the direction of $\boldsymbol{a}$ changes while the component perpendicular to $\boldsymbol{a}$ remains constant. If this latter component is zero initially, then it remains zero throughout the motion and there is no motion normal to $\boldsymbol{a}$; i.e. the particle travels along a straight line (along or opposite to the direction of $\boldsymbol{a}$). If the component of initial velocity normal to $\boldsymbol{a}$ is non-zero, then the trajectory is a parabola.

For example, a stone thrown without any horizontal velocity moves along a vertical line (along or opposite to the direction of gravity), but a stone thrown at an angle travels along a parabola.

S87 *Well then, for motion with variable* $\boldsymbol{a}$ *is the trajectory of the particle necessarily curved?*

T87 No. A particle may have varying $\boldsymbol{a}$ and yet can travel along a line in the same direction. For instance, a vehicle moving along a straight highway may accelerate at one time and decelerate at another time, which means its acceleration is not uniform even though the motion is along a straight line.

Another familiar example is a linear simple harmonic motion, in which acceleration varies in magnitude as well as direction.

S88 *What can you say about the acceleration vector if motion of a particle is given to be along a straight line?*

T88 All that you can conclude is that the components of acceleration transverse to the straight line are zero. (If they were not zero the particle would acquire transverse velocity components.) Nothing can be concluded about the acceleration component along the line. This component may or may not vary in magnitude and sign.

Displacement and path length

S89 For motion in space, is the magnitude of the displacement vector equal to the path length?

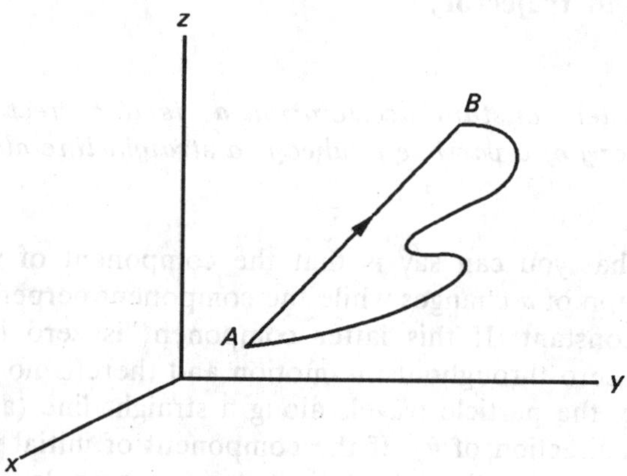

Fig. 3.7 The trajectory of a particle in space.

T89 No, not in general. As shown in Fig. 3.7, a particle may travel from a point A to a point B along any curved path. The displacement vector from A to B is along the straight line joining the two points; the magnitude of displacement is just the length of the line AB. The path length is the length of the actual curve followed by the particle. Obviously, therefore, the magnitude of the displacement is less than or equal to the path length. This equality holds only when the particle travels along the straight line AB without any turning back during its motion.

S90 *Why did you use the last phrase: without turning back?*

T90 The particle may travel along a line, but if it changes its direction back and forth then the path length is greater than the length of the line AB. This is why, even for one-dimensional motion, one must distinguish between the path length and the magnitude of displacement. Only when the motion is along a straight line in a **fixed** direction **is** the path length equal to the magnitude of the **displacement vector**.

Average speed and average velocity

S91 *For two or three-dimensional motion, is the average speed the same as the magnitude of average velocity?*

T91 No. As we have seen this is not even true in one dimension. The average speed is given by the total path length divided by the time required to travel the path. On the other hand, for average velocity we are not concerned with the actual path. It is just the displacement vector from the initial to the final position that is needed.

Now, the magnitude of displacement vector is either less than or equal to the total path length, as already explained. Hence average speed is not the same thing as the magnitude of average velocity. Only when motion is along a straight line in a fixed direction are the two quantities equal. In all other cases, the average speed is greater than the magnitude of average velocity.

S92 *What is the relation between instantaneous speed and the magnitude of instantaneous velocity?*

T92 They are equal to each other.

S93 *How is it that instantaneous speed = magnitude of instantaneous velocity, but average speed ≥ magnitude of average velocity?*

T93 For the same reason that I explained for one-dimensional case. Definitions of instantaneous quantities need an infinitesimal path or displacement. Now, the length of a finite portion of a curve is greater than the length of the chord (See Fig. 3.7) but on an infinitesimal portion the difference between the two is vanishingly small. Hence the result.

Relative velocity

S94 *Another question. Often some problems in mechanics involve the determination of relative velocity, which amounts to subtracting two velocity vectors. Can you explain the rationale behind it?*

T94 Let us think in terms of concrete problems.

Suppose it is raining on a windy day. The downward velocity of raindrops is 50 ms^{-1} and the wind velocity is 10 ms^{-1} in the east to west direction. What is the direction in which a boy standing on the ground should hold his umbrella?

Now, suppose the wind stops blowing and the boy rides on a bicycle with the same speed, i.e. 10 ms^{-1} and in the same direction i.e. east to west. What is the direction in which he should now hold his umbrella?

S95 *I suppose the answers are the same. Because the velocity of the wind in the first problem and that of the boy in the second problem are identical.*

T95 No. This is something you should be very clear about. In the first problem two velocities are given in a frame and we need to find the resultant velocity in the same frame. As seen from Fig. 3.8 (a), the boy should hold his umbrella at an angle $\tan^{-1}(1/5)$ with the vertical towards east. This problem is not a problem of relative motion.

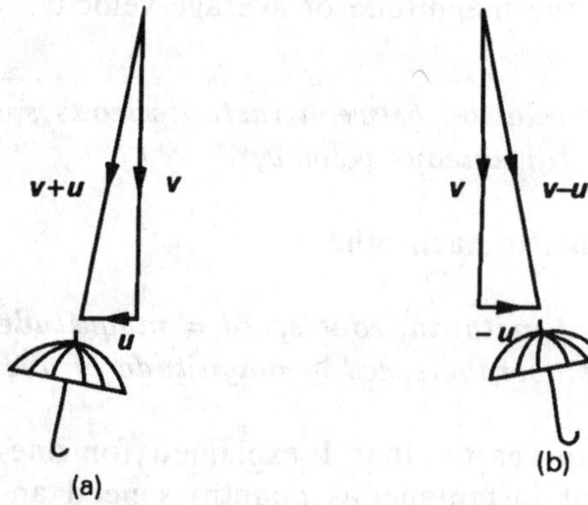

Fig. 3.8 (a) Resultant velocity of raindrops in the ground frame. (b) Velocity of raindrops relative to the bicyclist's frame.

In the second situation the velocity of rain in the ground frame is given and what is needed is the velocity of the rain in the frame of reference of the bicyclist. This, therefore, is a problem of relating the velocities of the same object as viewed from two different frames in relative motion.

S96 *And how does one relate them?*

T96 Suppose an object has a velocity $\boldsymbol{v}$ relative to a frame S and a velocity $\boldsymbol{v}'$ relative to frame S'. S' is moving with a velocity $\boldsymbol{u}$ with respect to S. From Box 3.2, it is seen that $\boldsymbol{v}' = \boldsymbol{v} - \boldsymbol{u}$.

Now here $\boldsymbol{v}$ is 50 ms^{-1} vertically downwards and $\boldsymbol{u}$ is 10 ms^{-1} east to west. Subtracting $\boldsymbol{u}$ from $\boldsymbol{v}$ we get $\boldsymbol{v}'$ as shown in Fig. 3.8(b). In other words the boy should hold the umbrella at an angle $\tan^{-1}(1/5)$ with the vertical towards the west.

S97 *That is now clear. Just one aside. We have been using the word 'particle' quite often. I should have asked this before. Is a particle the same thing as a body?*

T97 A particle is an idealization. You might consider it a body of negligible, ideally zero, size.

Box 3.2

Let S be a certain frame of reference and S' be another one moving with uniform velocity $\boldsymbol{u}$ with respect to S (See Fig. 3.9). $\boldsymbol{R}$ is the position vector of O' (the origin of S' frame) with respect to

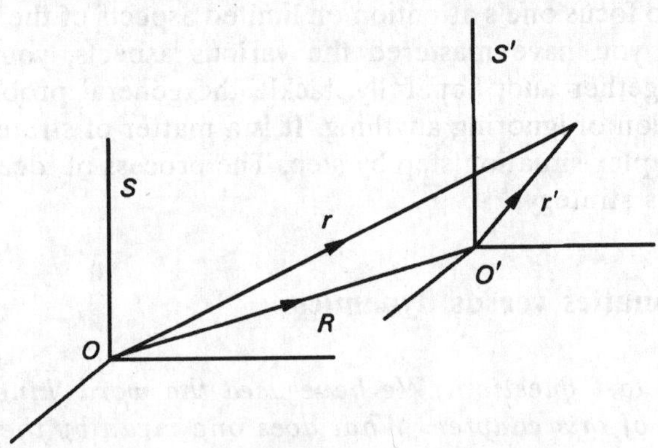

Fig. 3.9 Two frames S and S' in relative motion.

the S frame. $\boldsymbol{r}$ and $\boldsymbol{r}'$ denote the position vectors of a particle with respect to the S and the S' frames, respectively. It is clear from the figure that

$$\boldsymbol{r} = \boldsymbol{r}' + \boldsymbol{R}.$$

We differentiate this with respect to time and identify $d\boldsymbol{R}/dt = \boldsymbol{u}$, $d\boldsymbol{r}/dt = \boldsymbol{v}$ and $d\boldsymbol{r}'/dt = \boldsymbol{v}'$ to obtain the velocity addition rule

$$\boldsymbol{v} = \boldsymbol{v}' + \boldsymbol{u}.$$

S98 *O.K. but what is the point in this idealization?*

T98 You see, we are focussing here on one type of motion, the so-called translational motion. We are ignoring all other modes of motions that a body might have: rotation, vibration, etc. It is, therefore, best to consider an idealized body which does not have these latter modes of motion but has only translational motion. A point object, i.e. a particle, cannot possess any other type of motion except translation.†

S99 *What is the use of considering an idealized body that does not exist, and ignoring the realistic aspects of motion?*

T99 This is a rather general question. It really comes down to the basic strategy of doing physics, or at least learning physics. A realistic problem such as the general motion of an arbitrary body is complicated by many features. If you try to address all the aspects right from the beginning, life will become complicated and chances are that you will not get anywhere. It is a good practice, therefore, to truncate the problem; make simplifying idealizations which enable one to focus one's attention on limited aspects of the problem. Later, when you have mastered the various aspects, you can put them all together and, hopefully, tackle the general problem. It is not a question of ignoring anything. It is a matter of strategy for handling a complex situation step by step. The process of idealization is basic to this strategy.

Kinematics versus dynamics

S100 *One last question. We have used the word 'kinematics' in the title of this chapter. What does one mean by the term?*

T100 Kinematics deals with quantities that characterize motion: displacement, velocity, acceleration etc.; the relations between them, and their transformation properties from one frame to another—things which follow from the very definitions of these quantities. We know that accelerations are caused by forces but the investigation of these causes is not our concern here. When we examine the causes of accelerations in a problem we enter the domain of dynamics.

† Note, the oscillatory motion of a point particle as also its rotational motion are instances of its translational motion. It is wrong to associate these motions with vibrational and rotational degrees of freedom. The latter can arise only for an extended body.

S101 *I have heard something like that before, but can you explain this point through some examples?*

T101 The kinematic relations in T35 or more generally in T54, for example, follow from the definitions of velocity and acceleration. They have nothing to do with the laws of motion which relate force and acceleration. If these laws were different, the kinematic relations would nonetheless be the same. For instance, the laws of motion for inertial observers are different from those for non-inertial observers. But given acceleration in a frame, the same kinematic relations hold good whatever the frame may be.

As another example, the Galilean rule of velocity addition that we used in T96 is a kinematic relation, not a dynamical law.

Yet another example of kinematic relations is the Lorentz transformation equations of special relativity. These relate positions and times in one frame to those in another. They follow directly from the operational definitions of these quantities and the postulate of the constancy of velocity of light.

On the other hand, the relation

$$\boldsymbol{F} = \frac{d\boldsymbol{p}}{dt} \quad \text{where } \boldsymbol{p} = \frac{m_0\boldsymbol{v}}{\sqrt{1 - \frac{v^2}{c^2}}}$$

is a dynamical law of relativistic mechanics. In the same way, Maxwell's equations are dynamical laws of electromagnetism. So is the Lorentz force equation for a particle of charge q in an electromagnetic field,

$$\frac{d\boldsymbol{p}}{dt} = q(\boldsymbol{E} + \boldsymbol{v} \times \boldsymbol{B})$$

To sum up, kinematics involves merely the definitions of the quantities characterizing motion and the relations that follow from these definitions. Dynamics involves knowing the nature of forces and interactions and how they cause motion. Dynamics, as they say, is real physics.

S102 *I am waiting to plunge into it.*

4

Laws of Motion

Laws and definitions

S1 Newton's I law states that if there is no force acting on it, a body at rest remains at rest, and a body in uniform motion keeps moving with uniform velocity. Isn't this contrary to experience? We do seem to require a force to keep a body moving.

T1 Your observation is right. Indeed the ideas of Aristotle in this regard seem to be more in accord with our ordinary experience. He said that rest is the 'natural' state of a body and every body tends to achieve this state. Motion is unnatural and you need a force to keep a body in motion.

But now we know this to be completely wrong.

S2 *What is wrong with it?*

T2 You see, in practice, there are always forces which tend to oppose motion (friction, air resistance etc.). If these forces are made smaller and smaller, motion is less and less impeded. So in the ideal case of no friction, uniform motion can continue indefinitely without any external force.

It was Galileo who arrived at this fundamental insight: no cause is required to keep a body in uniform motion along a straight line (i.e. with constant $\boldsymbol{v}$). Newton's I law is a formal statement of just this insight.

S3 *Is the I law of motion true universally?*

T3 You must appreciate that the laws of physics relate to physical quantities measured relative to some observer (or frame of reference)†.

To answer your question, the I law of motion is true relative to certain frames of reference (called inertial frames) and not true relative to others (non-inertial frames).

† The notion of 'frame of reference' is discussed in detail in Chapter 5.

S4 *And how does one decide which frame is inertial and which non-inertial?*

T4 There is no *a priori* way of deciding that. The only way to decide whether a frame is inertial or not is to test the validity of Newton's I law. If the law is found empirically to be true, the frame is inertial. If not, the frame is non-inertial.

S5 *This puzzles me. First you say Newton's I law is valid in inertial frames, and then you define an inertial frame to be the one in which the I law is valid.*

T5 Well, it is best to view Newton's I law as a criterion by which one can select, from the set of all frames of reference, a special subset called inertial frames. In other words, Newton's I law only defines an inertial frame.

S6 *But I could use some different criterion for defining some other special class of frames? What is so significant about inertial frames?*

T6 Inertial frames are significant because the laws of motion take their simplest form in these frames. Newton's I law, as you will agree, is a rather simple criterion. It says that for the simple situation of no force, motion is also simple (namely uniform motion). Likewise, the II law of motion takes its simplest form in an inertial frame. In non-inertial frames it gets complicated by additional effects as you will see later.

S7 *You said that to see whether a frame is inertial or not, I should check the I law, i.e. test whether motion of a body is uniform in the absence of forces. But what does one mean by a force? How can I be sure that this as yet undefined thing is zero?*

T7 You must appreciate that at the very fundamental level physical reasoning is inevitably circular. You are right; the term force is yet to be defined properly, and still we are using it in the criterion for an inertial frame. But we do have some intuitive idea about this quantity. Force is some kind of influence on a body. We know by experience that this influence comes either by contact with other bodies or by non-contact effects such as gravitation or electromagnetism.† Thus, to ensure that force is absent, we should see that the body is not under any such external influences.

---

† The fact that forces such as friction, reactions, tension etc. are also electromagnetic in origin at the microscopic level is irrelevant to our discussion. In mechanics, these forces are treated as phenomenological forces different from gravitational, electric and magnetic forces; and this is the practice we shall adopt throughout.

For example, a body far out in space, far from all gravitating objects, may be said to be in a no-force environment. Now if this body has uniform motion relative to a frame, the frame is inertial.

S8 *That looks like an impractical criterion to use. Can't one judge the inertial nature of a frame, say the laboratory frame, by terrestrial experiments where we cannot get rid of gravity?*

T8 When it is impossible to create a no-force environment, the proper way to ascertain whether a frame is inertial or not, is to apply the II law to the motion of a test particle. If the observed acceleration of the test particle is accountable in terms of the material forces† acting on the particle in accordance with the II law, we conclude that the frame is inertial.

S9 *In an inertial frame Newton's II law states that* $\mathbf{F} = m\mathbf{a}$. *The only symbol in this equation that I understand clearly is acceleration* $\mathbf{a}$. *The symbol* $\mathbf{F}$ *denòting 'force' has not been defined independently in a quantitative manner, though we do have some intuitive idea about it. In the same way the symbol m denoting 'mass' has not been defined separately. How can an equation containing two undefined quantities become the most basic law of physics?*

T9 First of all recall what I said in T7. The second law is a classic example of how, at the most fundamental level, definitions, laws and measurements merge into one circular but mutually consistent scheme. Let us understand this better:

Start with an intuitive notion of force as some kind of push. Take two bodies 1 and 2, and let body 2 be made of two identical bodies of type 1. Use the same force and measure accelerations of 1 and 2 in turn. (That the force is the same in the two cases is to be judged intuitively.) It turns out that

$$\frac{a_2}{a_1} = \frac{1}{2}.$$

Now, it is reasonable to say that body 2 contains twice as much

† The reader should note the terminology that we shall use throughout. We call forces which are attributable to some physical agencies as material forces. (Examples: gravitational force, electric and magnetic forces, contact forces, etc.) These are to be distinguished from pseudo-forces (also called inertial forces), which have kinematic origin, as explained in Chapter 5.

matter as body 1. If we denote the 'quantity of matter' in a body by m, and call it the mass of the body, then

$$\frac{m_2}{m_1} = 2.$$

Conclusion: For a fixed force, $m_1 a_1 = m_2 a_2$ = constant.

We now make a jump and take this equation to be true for any pair of bodies. This, then, enables us to have an operational procedure for determining masses of different bodies: keep the force fixed and obtain ratios of masses from the measured values of accelerations. Assigning some arbitrary value of mass to any one body, the masses of all other bodies then can be determined.

We thus arrive at the important notion that to each body we can assign a property called mass. It is additive in character (i.e. if two bodies have masses m_1 and m_2, then their composite has a mass $m_1 + m_2$) and it determines the response of the body to a force. The greater the mass, the lesser the acceleration for a given force. Lesser acceleration means lesser change in the state of motion. The property of a body to resist change in its state of motion is called inertia. The constant appearing in Newton's II law of motion is, therefore, called inertial mass.

S10 *I see the way you arrived at an operational definition of mass, using the intuitive notion of a fixed force. But I must also have a way of obtaining the measure of force* $\mathbf{F}$ *independently. How. otherwise, will I test the validity of the equation* $\mathbf{F} = m\,\mathbf{a}$*?*

T10 You must get this right. You see, the II law *defines* the force $\boldsymbol{F}$ to be equal to $m\boldsymbol{a}$. There is no meaning in checking this equation as it is. $\boldsymbol{F} = m\boldsymbol{a}$ is true by definition.

S11 *That baffles me. How can the most basic law of nature be a mere definition, which, after all, is our construct? That way, I can concoct my own definitions. For example, I might define force by the relation* $\mathbf{G} = ma^2\mathbf{a}$*. What makes one definition more correct than the other?*

T11 Recall our discussion in T23 of chapter 2. No definition is right or wrong. A definition can be natural, significant and very useful. Or it can be contrived, trite and useless. The Newtonian 'definition of force has been found to be incomparably useful. In contrast, your definition is not wrong; it is useless.

S12 *On what grounds do you say that?*

T12 The great significance of the Newtonian definition of force is due to the following reasons:

First, it is the simplest quantitative definition of force that is consistent with the I law. When $\boldsymbol{F} = 0$, $\boldsymbol{F} = m\boldsymbol{a}$ implies $\boldsymbol{a} = 0$. Your definition of force in S11 is also consistent with the I law, but you will agree that it is not as simple.

Second, the crucial thing about the Newtonian definition is that simple and universal forms for $\boldsymbol{F}$ exist in nature. You must be familiar, for example, with the inverse square form of the gravitational force between two masses and the electrostatic force between two charges. The equation $\boldsymbol{F} = m\,\boldsymbol{a}$, as it stands, is just a definition; but combined with a definite form for a force, it yields an equation of motion for a particle under the action of the force. This equation is far from a mere definition. It is a law of nature with verifiable consequences. Other definitions of force that one might concoct will not admit of such simple forms in nature.

Lastly, an equally important thing about the Newtonian definition is that it satisfies the III law: the force $\boldsymbol{F}_{12}$ on particle 1 by particle 2 is equal and opposite to the force $\boldsymbol{F}_{21}$ on particle 2 by particle 1, i.e. $\boldsymbol{F}_{12} = -\boldsymbol{F}_{21}$. The definition $\boldsymbol{F} = m\boldsymbol{a}$ combined with the third law leads to the law of conservation of momentum of an isolated system. Once again, your definition of force stands nowhere in comparison. It does not satisfy the III law for $\boldsymbol{G}$, i.e. $\boldsymbol{G}_{12} = -\boldsymbol{G}_{21}$ is <u>not</u> true in nature.

It is these three statements which convert $\boldsymbol{F} = m\,\boldsymbol{a}$ from a mere definition into one of the most basic laws of physics. To sum up, the greatness of Newton lay in discovering a definition of force which admitted several simple and universal features. It is these features that underlie the enormous predictive power of Newtonian mechanics.

Newton's laws for extended bodies

S13 *Let me turn to some other queries. Are Newton's laws to be applied to point particles or are they applicable to extended bodies, i.e. bodies of finite size also?*

T13 Strictly, the law $\boldsymbol{F} = m\boldsymbol{a}$ is to be applied to a point particle. $\boldsymbol{F}$ here stands for the total force on the particle by the rest of the universe. In the same way, the III law is a statement on the mutual force between two point particles 1 and 2 $\boldsymbol{F}_{12} = -\boldsymbol{F}_{21}$.

S14 *Then, how come we often apply these same laws to extended bodies?*

T14 An extended body is made up of a number of particles. By applying Newton's laws to the constituent particles we can arrive at the laws for systems of particles. It turns out that for an extended body (or more generally for a system of particles), we obtain the equation $\boldsymbol{F} = m\boldsymbol{a}$, where $\boldsymbol{F}$ stands for the net external force on the body, m is the mass of the body and $\boldsymbol{a}$ is the acceleration of the centre of mass (CM) of the body. To avoid any confusion, it is better to write it as

$$\boldsymbol{F}^{(ext)} = m\boldsymbol{a}_{CM}. \tag{1}$$

This equation does not follow from the II law alone; we also need to use the III law. Similarly, if $\boldsymbol{F}_{AB}$ is the force on body A by body B, and $\boldsymbol{F}_{BA}$ is the force on body B by body A, then it can be seen, using Newton's laws, that $\boldsymbol{F}_{AB} = -\boldsymbol{F}_{BA}$. This is the III law for forces between extended bodies. (See Box 4.1.)

S15 *Is the II law $\boldsymbol{F} = m\boldsymbol{a}$ universal?*

T15 If by $\boldsymbol{F}$ in this equation we mean only the material forces, i.e. forces arising from physical agencies (e.g., gravitational force, electromagnetic force, contact forces like friction etc.), then the II law is true only in an inertial frame. However, the same equation is applicable even in a non-inertial frame, provided we include in $\boldsymbol{F}$ both material forces as well as suitable frame-dependent forces (called pseudo-forces). We shall talk about this in chapter 5.

Box 4.1

Consider a system of N particles. The II law, applied to the ith particle, yields

$$\begin{aligned} m_i\boldsymbol{a}_i &= \boldsymbol{F}_i \\ &= \boldsymbol{F}_i^{(ext)} + \sum_{\substack{j=1 \\ j\neq i}}^{N} \boldsymbol{F}_{ij}, \end{aligned} \tag{2}$$

where F_{ij} is the force on the i^{ih} particle by the jth particle, so that the second term on the right hand side is the net force on the ith particle due to all the remaining particles of the system. $F_i^{(ext)}$ is the net external force on the ith particle. Summing over all the N particles of the system, we get

$$\sum_{i=1}^{N} m_i \boldsymbol{a}_i = \sum_{i=1}^{N} \boldsymbol{F}_i^{(\text{ext})} + \sum_{\substack{i,j=1 \\ i\neq j}}^{N} \boldsymbol{F}_{ij}.$$

The second term on the right hand side contains pairs of terms like F_{12}, F_{21}, F_{45}, F_{54}, etc. Each pair sums upto a null force because of the III law. Therefore,

$$\sum_{i=1}^{N} m_i \boldsymbol{a}_i = \boldsymbol{F}^{(ext)}, \tag{3}$$

where $F^{(ext)}$ is the vector sum of all the external forces on the different particles of the system.
Now, the CM of the system is defined by

$$\boldsymbol{R} = \frac{\sum_{i=1}^{N} m_i \boldsymbol{r}_i}{\sum_{i=1}^{N} m_i} = \frac{\sum_{i=1}^{N} m_i \boldsymbol{r}_i}{m}$$

where m is the total mass of the system. This gives

$$\sum_{i=1}^{N} m_i \frac{d^2 \boldsymbol{r}_i}{dt^2} = m \frac{d^2 \boldsymbol{R}}{dt^2} \quad \text{i.e.} \quad \sum_{i=1}^{N} m_i \boldsymbol{a}_i = m\boldsymbol{a}_{CM}$$

With this, Eq.(3) takes the form

$$m\boldsymbol{a}_{CM} = \boldsymbol{F}^{(ext)}$$

which is Newton's II law for an extended body.

Consider next the mutual force between two extended bodies A&B. The force F_{AB} on body A by body B is given by

$$\boldsymbol{F}_{AB} = \sum_{a=1}^{N} \sum_{b=1}^{M} \boldsymbol{F}_{ab}$$

where the index a runs over all the N particles which constitute body A, and the index b runs over all the M particles which constitute body B. The force $\boldsymbol{F}_{BA}$ on body B by body A is given, similarly, by

$$\boldsymbol{F}_{BA} = \sum_{a=1}^{N} \sum_{b=1}^{M} \boldsymbol{F}_{ba}.$$

Since the individual forces are equal and opposite,

$$\boldsymbol{F}_{AB} = -\boldsymbol{F}_{BA},$$

which is Newton's III law for extended bodies.

Variable mass systems

S16 *Well. then, in an inertial frame is the relation* $\boldsymbol{F} = m\boldsymbol{a}$ *true in all situations?*

T16 In non-relativistic mechanics, where we usually deal with well-defined systems with a fixed mass, the II law in the above form is always valid. But, then, there are some systems, which have a variable mass. For example, a raindrop falling under gravity and collecting mist on its way has a mass that grows with time. Similarly, a rocket that propels itself forward by ejecting its combustion products has a mass that decreases with time. In these situations the II law must be seen as an instantaneous relation

$$\boldsymbol{F}(t) = m(t)\boldsymbol{a}(t) ,$$

where $m(t)$ is the instantaneous mass of the system. Thus, even when the system we are dealing with is a variable mass system, we apply the II law to a system of definite mass: the one which has mass m at time t. $\boldsymbol{F}$, therefore, stands for <u>all</u> the forces acting on this system at that instant. In particular, it must include the force exerted by the mass entering or leaving the system. This will be clear from Box 4.2.

Mass, even of a definite system, is not a strict constant. According to Einstein's theory of relativity, the mass of an object increases with its speed. At ordinary speeds the mass variation is negligible, but for speeds comparable to that of light the mass variation is appreciable and modification of Newtonian law of motion is necessary. (See chapter 5.)

Box 4.2

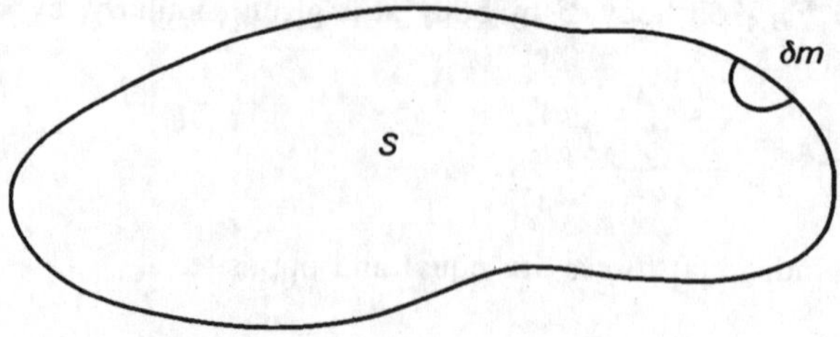

Fig. 4.1 Schematic representation of a variable mass system S of mass m at time t. The element of mass δm enters/leaves the system in time δt.

Consider a system with variable mass, shown schematically in Fig. 4.1. At time t, the mass of the system is m. During the small time interval δt, a small mass δm enters or leaves the system so that the mass at time $t + \delta t$ is $m + \delta m$. We apply the II law to the system at time t. Thus the system to which we are applying the II law is S whose mass is m. The net force responsible for the acceleration of the CM of the system S includes, in addition to all the other external forces, the force exerted on S by δm since the mass element δm is external to the system S. Newton's II law gives:

$$m\frac{d\boldsymbol{v}}{dt} = \boldsymbol{F}_m^{\text{ext}} + \boldsymbol{F}_{m,\delta m}, \tag{4}$$

where $\boldsymbol{F}_m^{\text{ext}}$ is the net external force on S and $\boldsymbol{F}_{m,\delta m}$ is the force on S by δm. By Newton's III law

$$\boldsymbol{F}_{m,\delta m} = -\boldsymbol{F}_{\delta m,m}. \tag{5}$$

Now, if the velocity of δm at time t is $\boldsymbol{v}_b$, and that at $t + \delta t$ is $\boldsymbol{v}_a$, then the rate of change of momentum of δm is

$$\left(\frac{d\boldsymbol{p}}{dt}\right)_{\delta m} = \lim_{\delta t \to 0} \frac{|\delta m|(\boldsymbol{v}_a - \boldsymbol{v}_b)}{\delta t}.$$

If the change in velocity $(\boldsymbol{v}_a - \boldsymbol{v}_b)$ is infinitesimal, then $(d\boldsymbol{p}/dt)_{\delta m} = 0$. If, however, in the process of entering or leaving the system, the element δm undergoes a finite change in velocity (which is possible if the force on δm is impulsive), then

$$\left(\frac{d\boldsymbol{p}}{dt}\right)_{\delta m} = \left|\frac{dm}{dt}\right| (\boldsymbol{v}_a - \boldsymbol{v}_b) \tag{6}$$

Now, since $(d\boldsymbol{p}/dt)_{\delta m} = \boldsymbol{F}^{\text{ext}}_{\delta m} + \boldsymbol{F}_{\delta m,m}$ we can use Eqs. (5) and (6) to express Eq. (4) in the form

$$m\frac{d\boldsymbol{v}}{dt} = \boldsymbol{F}^{\text{ext}}_{m} + \boldsymbol{F}^{\text{ext}}_{\delta m} - \left|\frac{dm}{dt}\right| (\boldsymbol{v}_a - \boldsymbol{v}_b)$$

which is the equation of motion for a variable mass system. The force $\boldsymbol{F}^{\text{ext}}_{\delta m}$ on the small mass element is usually negligible[†] and, hence, we write the equation in the form

$$m\frac{d\boldsymbol{v}}{dt} = F^{\text{ext}}_{m} - \left|\frac{dm}{dt}\right| (\boldsymbol{v}_a - \boldsymbol{v}_b)$$

Consider the example of a falling raindrop. The mist droplets floating about in the air can be considered to have zero velocity. When a mist droplet sticks to a raindrop, its velocity changes instantaneously to that of the raindrop. Thus, for the mist droplet $\boldsymbol{v}_b = 0$ and $\boldsymbol{v}_a = \boldsymbol{v}$. The equation for this variable mass system can, therefore, be written as

$$\frac{d}{dt}(\mathrm{m}\,\boldsymbol{v}) = \boldsymbol{F}^{\text{ext}}.$$

In the case of rocket motion, fuel particles move with the rocket, so that $\boldsymbol{v}_b = \boldsymbol{v}$. After combustion they are ejected with a certain velocity; hence, $\boldsymbol{v}_a = \boldsymbol{v}'$. Thus, the equation of motion for a rocket is

$$m\frac{d\boldsymbol{v}}{dt} = \boldsymbol{F}^{\text{ext}} - \left|\frac{dm}{dt}\right| (\boldsymbol{v}' - \boldsymbol{v})$$

Note that $(\boldsymbol{v}' - \boldsymbol{v})$ is the velocity of the exhaust relative to the rocket. If $\boldsymbol{F}_{ext} = 0$, this equation is easily obtained by momentum conservation as given in many textbooks.

Newton's III law

S17 *Let us turn to the III law. I have a number of queries about it. First, if the action and reaction forces on a body are equal and opposite, why don't they cancel each other out and produce a null effect?*

[†] In general, however, this is not true. Consider a vertically held chain being dropped on the ground. Every element of the chain, upon hitting the ground, is instantaneously brought to rest by the normal force exerted by the ground. In this case, $(dp/dt)_{\delta m} = F^{\text{ext}}_{\delta m}$ and $F_{\delta m,m} = F_{m,\delta m} = 0$.

T17 You have misquoted the III law and, therefore, have arrived at a wrong conclusion. Go back to your words 'action and reaction forces on a body are equal and opposite'. The III law never says that. In the III law we never talk of forces on the same body; rather we are relating the force on body A by the body B to the force on body B by the body A. These forces are equal and opposite but they clearly act on different bodies.

Of course, if you consider the composite system $(A+B)$, $\boldsymbol{F}_{AB}$ and $\boldsymbol{F}_{BA}$ are internal forces of the system and they do cancel each other because of the III law. In fact for any system of particles, all the internal forces cancel pairwise because of the III law, as we saw in Box 4.1.

S18 *Can you illustrate the point?*

T18 A familiar example is the collision of two bodies A and B with no external forces. Let us consider body A as our system. It experiences force $\boldsymbol{F}_{AB}$ during collision. It obviously cannot cancel with $\boldsymbol{F}_{BA}$ since the latter acts on body B which is not a part of the system under consideration. Consequently, body A suffers a change in momentum. Likewise, if we consider body B as the system to which to apply Newton's laws, then, an external force $\boldsymbol{F}_{BA}$ acts on it during collision and the momentum of body B changes accordingly. We may, however, choose to apply Newton's laws to the composite system $(A+B)$. $\boldsymbol{F}_{AB}$ and $\boldsymbol{F}_{BA}$ are, then, internal forces acting on the <u>same</u> system and it makes sense to add them up. They produce a null result since they are equal and opposite by Newton's III law. There is no other external force on the system, and hence the total momentum of the composite system is conserved.

S19 *Let me ask a standard question. In a horse driven carriage, the horse pulls the cart with some force. By the III law the cart pulls the horse with the same force in the opposite direction. For the cart–horse system, therefore, these forces cancel out. What makes the carriage move?*

T19 What you have said is perfectly correct, and if the cart–horse system were far away in empty space, there would be no external force on the system and the momentum of the cart–horse system cannot change.

On the ground, however, the cart–horse system is not isolated. The horse pushes the ground in the backward direction at some angle. By the III law the ground exerts an equal and opposite force on the horse. The vertical component of this force, together with the normal force on the cart, balance the weight of the horse–cart system. The forward component of the force exerted on the horse by the ground causes the motion of the horse–cart system.

S20 *Isn't there a force of friction on the horse?*

T20 The horizontal force on the horse by the ground in the forward direction <u>is</u> the force of friction between the sole of the horse and the ground. It is what causes the horse to move. In the absence of this friction the horse cannot move.

S21 *Let me turn to another important doubt. In an action–reaction pair, does action precede reaction?*

T21 No doubt you have got this impression from the terms action and reaction; but it is not true. Remember, no causality whatsoever is implied in Newton's III law. It is wrong to think of action as the cause and reaction as its effect. Both the forces act at the same time, and any of the two can be called action or reaction.

S22 *I am not convinced. For instance, when I press a wall with some force, it is obvious that I am causing the force to act on the wall, and the reaction force by the wall on my hand is an effect or result of what I did.*

T22 You are both right and wrong. You are right because there is indeed a causality in what you have just said. But you are wrong in taking that for a causal relation between the action and reaction forces in the III law.

S23 *Can you please elaborate?*

T23 For something to be called a cause, it must precede effect. For example, suppose we hear a firecracker explode far away. In this case it is correct to say that the explosion of the firecracker is the cause and the reception of the sound is the effect, because clearly the cause comes before the effect.

But when I press a wall with a certain force, <u>instantaneously</u> an equal and opposite force by the wall acts on my hand; the two forces are simultaneous. So it makes no sense to call one 'the cause' and the other the effect'. †

---

† Here we are talking of contact forces. But in Newtonian mechanics even forces such as gravitational force and Coulomb force between bodies at a distance act instantaneously. This, of course, violates special relativity. (See chapter 5.)

S24 *But I did cause the force to act on the wall, didn't I?*

T24 True. You decided or intended to press the wall and as a result you did press the wall. There is what is sometimes called an 'intentional causality' here. Physics usually keeps away from analyzing intentional causality and the related question of free will etc.

S25 *That is irrelevant to my question. I could replace myself by a robot. The robot could be programmed to undergo a succession of steps, the end result of which is the pressing of the wall by the robot's hand.*

T25 The robot would have to be programmed by a human being, so ultimately you cannot get away from the question of intentional causality in this particular example. But suppose you restrict yourself to the robot and the wall. Then there does exist a causality between say the triggering of some switch and the robot hand eventually pressing against the wall. But this, by no means, implies any causal relation between the mutual forces between the robot hand and the wall, which occur simultaneously.

You must realise that the word 'action' in the III law simply stands for a force. It does not have anything to do with the causal succession of steps which lead to that force.

Basically, I think, the wrong notion that action precedes reaction has its origin in the way we use these words in ordinary language. An often misquoted example, albeit in a jocular vein, is: if you hit someone, he hits you back. These two events do not form an action–reaction pair in the sense of the III law. When you hit someone, the reaction, in the sense of the III law, is what you feel instantaneously in your hand. What the other person delivers later on is over and above this reaction force!

S26 *I have heard of the weak and strong form of the III law. What is it supposed to mean?*

T26 In the way that we stated the III law, the action and reaction forces are equal and opposite, but we have said nothing about their directions relative to the line joining the two particles. This is the weak form of the III law. In the strong form of the III law we say, in addition, that the action and reaction forces are along the line joining the two particles.

S27 *What is the significance of this difference?*

T27 When the mutual forces between a pair of particles satisfy the strong form of the III law, their lines of action coincide. Clearly, not only the net force but also the net torque on the pair, due to their mutual interaction, is zero. This means, the system by itself can have neither translational acceleration nor rotational acceleration.

On the other hand, if the forces of mutual interaction satisfy only the weak form of the III law, the net force on the system is zero but the net torque need not be zero.

S28 *Can you give some examples?*

T28 The Newtonian gravitational force between two masses, the electrostatic (Coulomb) force between two charges, the magnetic force between two infinitely long parallel currents are some examples of the strong form of the III law.

In contrast, the forces between two magnetic dipoles (or two steady current loops) satisfy only the weak form. Two magnetic dipoles left to themselves can produce a net torque around their centre of mass (See Fig. 4.2).

S29 *Is it correct to say that the III law, at least in the weak form, is valid universally?*

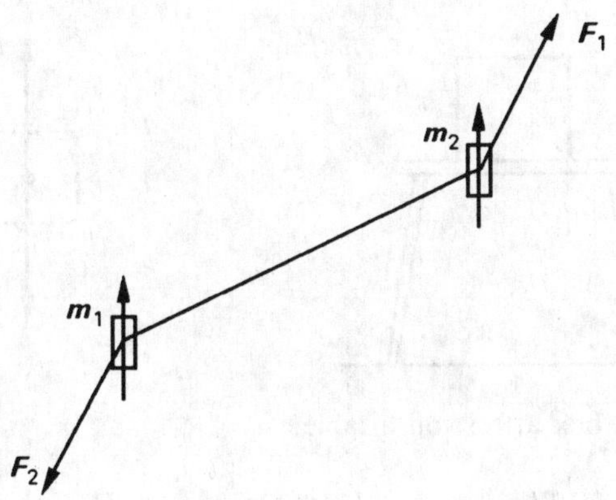

Fig. 4.2 The forces of mutual interaction between two magnetic dipoles do not act along the line joining them.

T29 No. The III law is not valid in the relativistic domain, i.e. for forces between charged particles in motion (See Chapter 5).

S30 *Another question. You said that in a non-inertial frame the II law is modified by the addition of pseudo-forces. Is the III law true in a non-inertial frame?*

T30 The III law refers to the mutual material forces between a pair of bodies A and B, and says that they are equal and opposite. In effect, it says that material forces in nature always occur in pairs. The III law for material forces is also true in a non-inertial frame. The pseudo-forces in a non-inertial frame, however, do not occur in pairs, and there is no question of a III law for them.

Illustrative problems

S31 *We have discussed so far some general points about the laws of motion. It will help if you take up some practice examples.*

T31 You are right. The best way to understand the laws of motion (indeed any part of physics) is to solve problems. But one must not equate problem-solving to learning some recipes. We should try to understand critically the principles used while solving a problem.

Let us begin with a very simple example. Consider a box at rest on a table, as shown in Fig. 4.3. How do you view the situation in terms of Newton's laws?

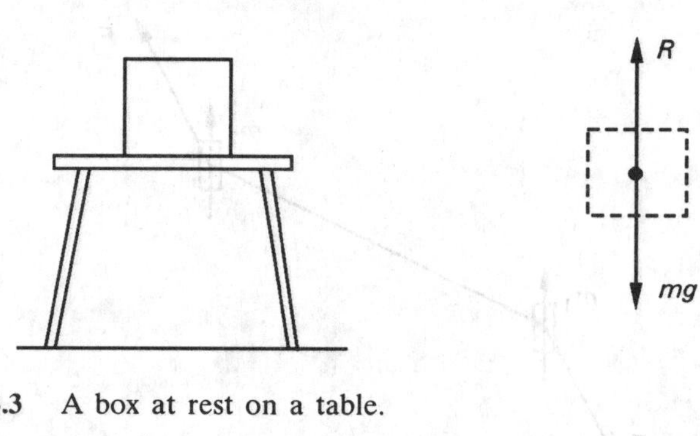

Fig. 4.3 A box at rest on a table.

S32 *That is easy. The weight of the box is* mg*. R is the reaction on the box by the table. Since action and reaction are equal and opposite* $R = mg$*, and thus the box is in equilibrium.*

T32 The explanation you have given is completely wrong. The forces R and mg, as shown in the figure, act on the same body. They cannot possibly be an action-reaction pair. The correct action–reaction pairs in the problem are:

gravitational force on the box by the earth = mg (downwards),
gravitational force on the earth by the box = mg (upwards)
and force on the table by the box = R (downwards),
force on the box by the table = R (upwards)

We see clearly that R and mg do not constitute an action–reaction pair.

S33 *Well, then, how do you explain the equilibrium of the box?*

T33 It is wrong to think that we can explain the equilibrium of the box. You see, the detailed origin of the force R is so complicated that one cannot determine its value *ab initio*. All we know about R is that it is a self-adjusting force up to the point the floor does not yield. Thus, if the box is <u>*given*</u> to be in equilibrium, we use this fact to conclude from the II law (not the III law) that $R = mg$.

S34 *That surprises me. Does it mean that R may not always be equal and opposite to mg?*

T34 That is right; it depends on the situation. Suppose, for example, the entire (table + box) assembly were in a lift going downwards with acceleration a. The action–reaction pairs in this case are the same as when the assembly was at rest. In particular, the action force on the table by the box is equal and opposite to the reaction force on the box by the table, since the III law is always true in mechanics no matter what the situation is. But since the box is not in equilibrium, $R \neq mg$. In fact, by the II law,

$$mg - R = ma.$$

This again shows how wrong it is to think of R and mg as an action–reaction pair.

S35 *I get your point. Any other example?*

T35 Consider the situation in Fig. 4.4. Two blocks of masses m_1 and m_2 are connected by an inextensible, massless string. The blocks are being pulled by a third block of mass m_3 via a frictionless pulley as shown in the figure. Determine the acceleration of the blocks.

S36 *This seems straightforward. The force on the blocks m_1 and m_2 is due the weight of the third block, i.e. m_3g. By the II law the acceleration is $m_3g/(m_1+m_2)$.*

T36 Your answer is wrong. You may avoid such errors, while applying the laws of motion, by proceeding systematically, step by step:

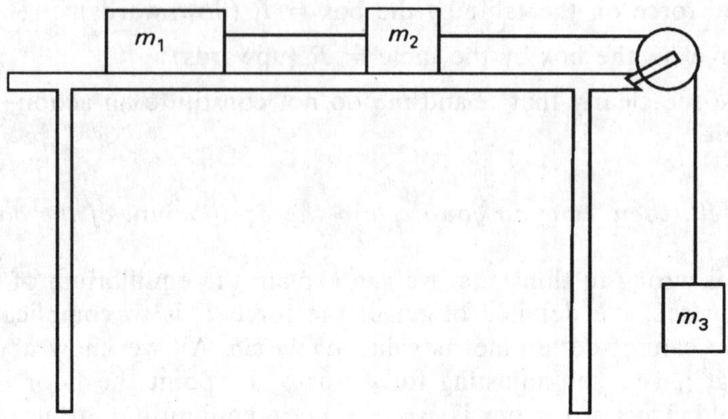

Fig. 4.4 Two masses m_1 and m_2 being pulled under the weight of a third mass m_3.

1) Identify clearly the system to which you are applying the laws. The system identified for the application of the II law is usually referred to as 'free-body'. In a given situation, one can take the entire system, or any part of it, as the a 'free-body'.†
2) Having identified a free-body, it is convenient to represent all the forces acting <u>on</u> the free-body in a diagram. Such a diagram is called a free-body diagram. In this diagram, never include the forces <u>by</u> the free-body on other parts of the system.
3) In mechanics, the possible forces on a body are the contact forces such as friction, reaction of a surface or tension in a string etc. and the non-contact force due to gravitation.
4) Sometimes you may need to analyze the situation by taking two or more freebodies one by one. In that case relate the appropriate forces using the III law.

S37 *Can you illustrate these points in the problem at hand?*

T37 Let us first take the mass m_3 as the free-body. The free-body diagram is shown in the Fig. 4.5(a). There are two forces acting on this free-body: the weight m_3g in the downward direction and tension T_2 in the upward direction. If we consider m_2 as the free-body, the forces are: weight m_2g downward, the normal reaction R_2 upward, tension T_2 to the right and tension T_1 to the left. Finally, for m_1,

† The term 'free-body' is a misnomer. Avoid a possible wrong impression that 'free-body' means a body on which the net force is zero.

the forces are: m_1g downward, the normal reaction R_1 upward and tension T_1 to the right. The free-body diagrams in these two cases are shown in Figs. 4.5(b) and 4.5(c) respectively.

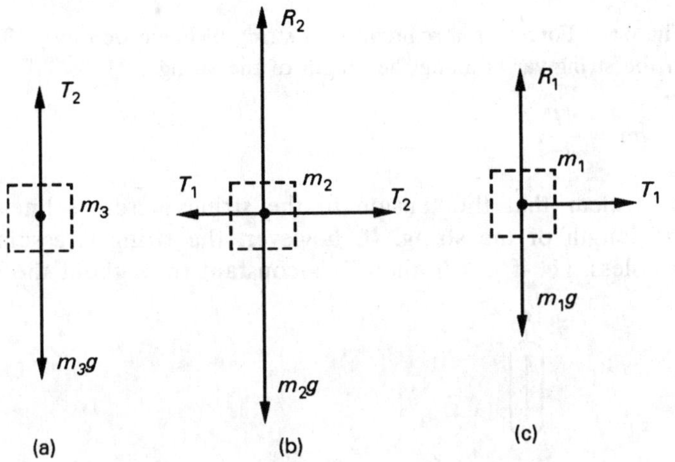

Fig. 4.5 Free-body diagrams for the three masses of Fig. 4.4.

As far as vertical direction is concerned, masses m_1 and m_2 are in equilibrium. Hence $R_1 = m_1g$ and $R_2 = m_2g$. Since the string is inextensible, the accelerations of the three masses must be equal in magnitude. We can therefore write the equations of motion for the three masses:

$$m_1a = T_1$$

$$m_2a = T_2 - T_1$$

$$m_3a = m_3g - T_2$$

These three equations in the three unknowns a, T_1 and T_2 can be easily solved to obtain $a = m_3g/(m_1 + m_2 + m_3)$.

S38 *One often takes the string to be massless. What if the string has a mass?*

T38 Consider a string of mass per unit length μ. A small element of length δx is subject to tension $T(x)$ on one side and $T(x+\delta x)$ on the other side, so that the net force on it is $(dT/dx)\delta x$ (See Fig. 4.6).

If a is the acceleration of the string, then

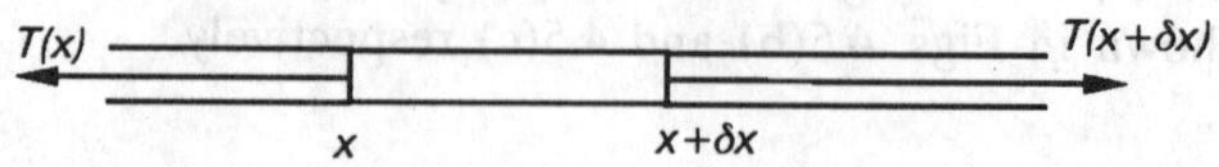

Fig. 4.6 Forces on a segment ot a string of linear density μ. The tension in the string varies along the length of the string.

$$\mu a = \frac{dT}{dx}.$$

It is clear that the tension in the string increases linearly along the length of the string. If, however, the string is assumed to be massless, i.e. if $\mu = 0$, then T is constant throughout the string.

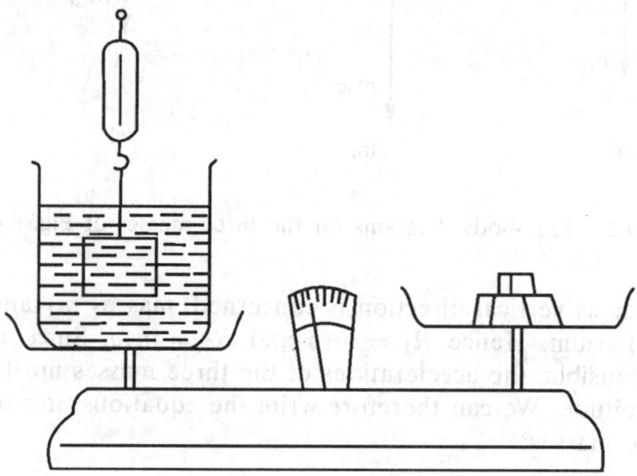

Fig. 4.7 The metal block, suspended from a spring balance, is immersed in water in the beaker. The beaker is then placed in one pan of a balance.

As the next example, consider the arrangement shown in Fig. 4.7. The weight of the beaker plus water is balanced by 1 kg of mass in the other pan. A piece of a metal of mass 0.08 kg and volume 10^{-5} m^3, suspended by a thread from a spring balance, is immersed in water. How much will the spring balance read? (Take $g = 10$ ms^{-2}.)

S39 *By Archimedes' principle, the upthrust on the metal piece, which causes a loss in its weight, is equal to the weight of the water displaced. The mass of the water displaced is 0.01 kg. The upthrust, therefore, is 0.1 N and hence the spring balance will read 0.7 N, i. e., 0.07 kg.*

T39 Fine. Now, tell me how much additional mass will I need to put in the other pan for balancing?

S40 *I do not know how to answer that.*

T40 Choose (beaker + water) as the free-body. Its mass is 1 kg. Now see the extra force on this free-body due to the immersion of the metal piece. Upthrust is the force on the metal piece by the water. By the III law, the metal piece exerts a downthrust on the water of the same magnitude. Thus the extra downward force on the free-body is 0.1 N. To balance this, we will need to put an extra 0.01 kg in the other pan.

As our last example, let me take up a very illustrative problem.

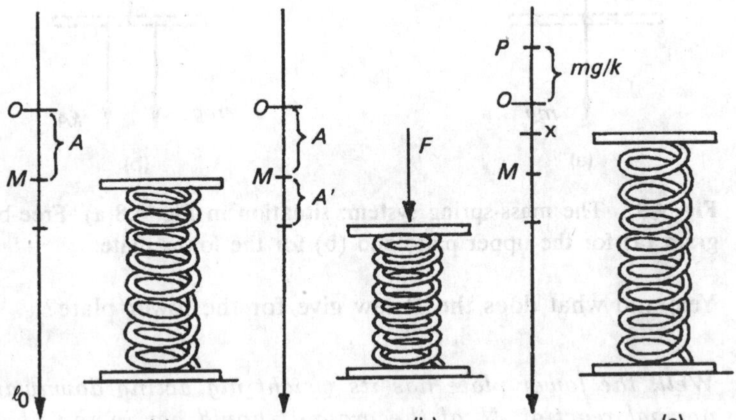

Fig. 4.8 The mass-spring system. (a) The spring is compressed under the weight of the upper plate. (b) The force F compresses the spring further. (c) An instantaneous configuration of the assembly after force F is released.

Suppose two plates each of mass m, connected by a massless spring, are kept on a table as shown in Fig. 4.8. The length of the spring in relaxed, the position is ℓ_0 and its stiffness constant is k. An external force F compresses the spring. Find the minimum compression of the spring necessary so that the assembly jumps when the force is removed.

S41 *How should I proceed?*

T41 First consider the assembly kept vertically without any external force as shown in Fig. 4.8(a). Let the compression of the spring in this position be A. How will you draw the free-body diagram for the upper plate?

S42 *The upper plate has only two forces acting on it: weight mg downward and the spring force—kA (upward). The free-body diagram for the upper plate is as shown in Fig. 4.9(a), and the II law gives*

$$mg - kA = 0. \tag{7}$$

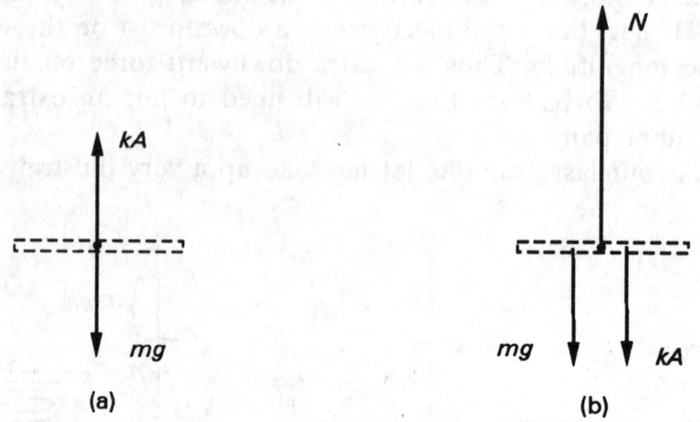

Fig. 4.9 The mass-spring system: situation in Fig. 4.8(a). Free-body diagram (a) for the upper plate and (b) for the lower plate.

T42 Yes; and what does the II law give for the lower plate?

S43 *Well, the lower plate has its weight mg acting downward. The normal reaction N of the ground should act in the upward direction: and the spring force on the lower plate, I believe, should act in the downward direction.*

T43 That is correct; the spring force is the mutual force between the two plates mediated by the spring. Hence the spring force on the lower plate must be equal and opposite to that on the upper plate.

S44 *Well then, the free-body diagram for the lower plate is as shown in Fig. 4.9(b) and the II law gives*

$$mg + kA - N = 0. \tag{8}$$

T44 Note that we can combine Eqs. (7) and (8) to obtain

$$N = 2mg$$

This is just what we should expect, since it describes the equilibrium of the system as a whole.

Now suppose an external force F is applied to the upper plate compressing the spring further through a distance A', so that the net compression of the spring is $A + A'$. (See Fig. 4.8(b)). What does the II law yield now?

S45 *The upper plate now has a force F acting downward in addition to the weight and the spring force. The free-body diagram for the upper plate, therefore, is as shown in Fig. 4.10(a) and the II law gives*

$$F + mg - k(A + A') = 0. \tag{9}$$

For the lower plate, the forces are the same as in the previous case.

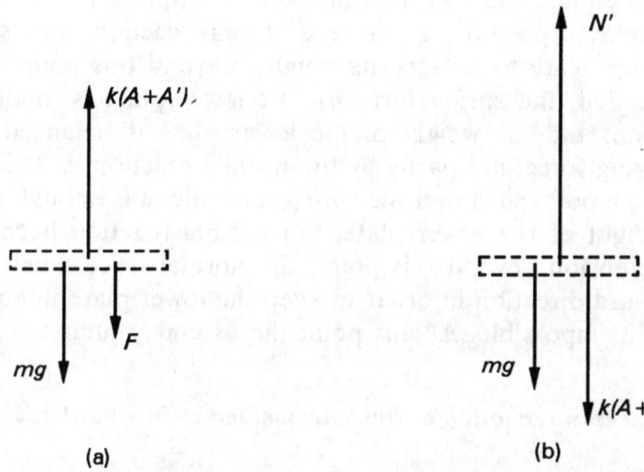

Fig. 4.10 The mass-spring system: situation in Fig. 4.8(b). Free-body diagram (a) for the upper plate and (b) for the lower plate.

T45 Yes, but now the normal reaction must be different.

S46 *Right, so the free-body diagram for the lower plate is as shown in Fig. (4.10b) and the II law gives*

$$mg + k(A + A') - N' = 0. \tag{10}$$

T46 Again notice that, combining Eqs. (9) and (10) we obtain

$$2mg + F - N' = 0.$$

This equation represents the equilibrium of the total system under the new conditions.

Now, when the force F is removed suddenly, the upper plate experiences a net unbalanced force of $mg - k(A + A')$. It is no longer in equilibrium; its acceleration at that instant is given by

$$mg - k(A + A') = ma.$$

The lower plate, on the other hand, has this self adjusting force of normal reaction acting on it. When the force F is removed, the magnitude of N' reduces instantaneously so that the next force on the lower plate is zero. The upper plate, however, is now in motion. As a result, the compression of the spring, and hence the spring force on the lower plate, varies with time. The normal reaction on the lower plate, therefore, must vary accordingly. As the upper plate moves upward, the compression of the spring decreases, and hence a smaller magnitude of the normal force is enough to keep the lower plate in equilibrium. At the time when the spring is instantaneously in its relaxed position, a non-zero normal reaction must still act on the lower plate to balance its weight. Beyond this point the spring is extended; the spring force on the lower plate is in the upward direction; and the weight of the lower plate is balanced partly by the spring force and partly by the normal reaction. If the extension reaches a point such that the spring force alone is enough to balance the weight of the lower plate, the normal reaction becomes zero. For extension beyond this point, the normal force must act in the downward direction in order to keep the lower plate in equilibrium, which is impossible. At this point the assembly jumps.

S47 *That seems reasonable, but can we see it in quantitative terms?*

T47 If we denote the instantaneous compression of the spring by x, then the equilibrium condition for the lower plate yields

$$N(x) = mg + kx$$

(See Fig. 4.8(c)). Note that distance x is measured from 0, taken as positive along the downward direction.) It is clear that the normal reaction becomes zero when the compression $x = -mg/k$ (Negative compression implies extension). For $x > -mg/k$, $N(x)$ is non-negative and the lower plate remains in equilibrium. For $x < -mg/k$, $N(x)$ is negative, which is impossible. Hence, at $x = mg/k$, i.e. when the upper plate reaches point P in Fig. 4.8(c), the assembly jumps.

Recall, however, that the force F compresses the spring through a distance of A'. The upper mass, therefore, performs oscillations with amplitude A' about the mean position M. So if A' is small enough ($A' < A + mg/k$), x will never be less than $-mg/k$, and the assembly will not jump. The minimum compression A' necessary for the spring to jump is, thus, $A + mg/k$.

The force with which the spring must be compressed, in order to produce this compression, is $k(A + mg/k)$. From Eq. (7) we have $kA = mg$. Hence the minimum force with which the spring must be compressed, so that it jumps when released, is $2mg$.

S48 *I appreciate the utility of the free-body technique for solving problems in mechanics. I suppose I need to solve many more problems to master it better.*

T48 That is right. Let us move on to something different.

5

Newtonian Relativity

Frames of reference

S1 *We use the term 'frame of reference' so often in physics. What does it mean precisely?*

T1 Any physical phenomenon is a succession of events. Its description depends on the state of motion of the observer. For example, the position of a box in a moving train remains constant relative to a person in the train, but changes in relation to someone on the ground.

For a given state of motion of the observer, the position co-ordinates of an event depend on the choice of the origin and orientation of the axes. The time of an event depends on the zero setting of the clock that one uses.

To summarize, any description of a physical phenomenon is always made relative to some observer in some particular state of motion, with a certain choice of origin, orientation of axes and the zero setting of a clock.

S2 *I understand that but you still have not explained the term 'frame of reference'.*

T2 Notice that I used words like 'relative to some observer'. The word 'observer', however, can be very misleading. It gives an impression that we are talking of a person looking at the phenomena, making appropriate measurements and possibly comparing them with those of another person. I suggest you banish this picture from your mind.

S3 *I am surprised. What is wrong with it?*

T3 Physics deals with numbers—measurements carried out by impersonal instruments. The person behind the instruments is irrelevant for physics; that is what one means by objectivity in science. So it is best to describe phenomena without invoking the notion of an observer, if that word connotes a subjective human observer.

S4 *Then how do you talk of the relativity of description of a phenomenon for different observers?*

T4 We replace the image of an 'observer' by an impersonal abstract notion of a 'frame of reference'. A 'frame of reference' is no real object; we simply *imagine* a frame of long rigid rods extending out from a point (origin) in space in three independent directions. This frame of reference is characterized by its state of motion relative to some material body of the universe. Thus, for example, the frame of reference of a train is an abstract artifact which has the same motion as that of the train.

Therefore, instead of saying, for example, that the trajectory of a stone dropped out of a running train is a straight line for a train observer and a parabola for a ground observer, it is better to say that the trajectory is a straight line in the train's frame of reference and a parabola in the ground's frame of reference.

S5 *That sounds like quibbling with words. Isn't that so?*

T5 In a sense it is. Physicists take the words 'observer' and 'frame of reference' to mean the same thing and often use them interchangeably. You must, however, take care not to have any human-like association with the word 'observer', and understand that 'observations' in physics mean measurements by physical instruments.

A lot of misconceptions about 'frames of reference' arise because one tends to go by the literal meaning of the word 'observer'.

Misconceptions about frames of reference

S6 *Can you warn me of these possible misconceptions about frames of reference?*

T6 First and foremost, you should never think of a frame of reference as a concrete physical object. A frame of reference is, as said earlier, characterized only by its state of motion. It does not have any other material object-like qualities. For example, when you are talking of a frame of reference of the earth, the frame is not to be equated to the earth. The earth's frame is an abstract artifact whose only relevant characteristic is its state of motion: it has the same motion as the earth, i.e. relative to this frame the earth is at rest. In addition, of course, you have to make some choice in this frame of the origin and orientation of the axes, and also of the zero setting of the clock used for measuring time. See Fig. 5.1.

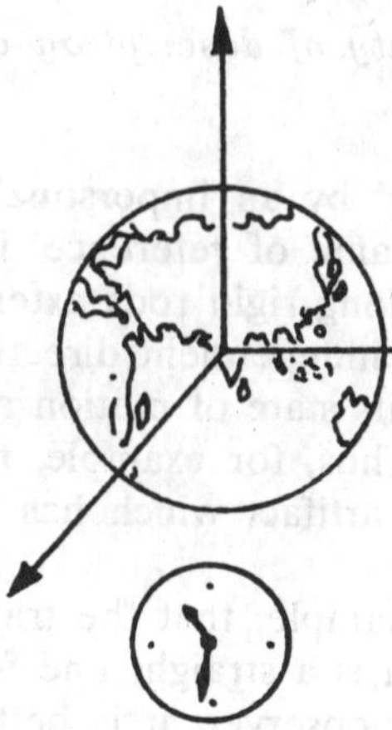

Fig. 5.1 A frame of reference of the earth.

S7 *Can the origin of a train's frame of reference be outside the train?*

T7 Of course it can be. That is just the pitfall I wanted you to avoid. You must clearly realize that, for the definition of the train's frame of reference, the size and other material characteristics of the train are totally irrelevant. It is wrong to think that the train's frame is 'bounded' by the extension of the train. The origin of this frame can be anywhere in space, and the orientation of its axes can be chosen arbitrarily.

S8 *But in a train's frame of reference, with its origin outside the frame, how can one look at the phenomena inside the train if its windows are shut?*

T8 Your statement is full of grave misconceptions about 'frames of reference'. You are thinking that there is literally a human observer who positions himself at the origin of the frame and makes observations by 'looking' at the phenomena. I have already said that in physics observing means measuring; and the process of measurement is not to be equated to viewing by humans. Measurement of any quantity in physics is carried out using some physical instrument. So whatever the origin and the axes of the train's frame of reference, measurements anywhere can be carried out by suitable instruments.

S9 *If a ball is thrown out of a running train, is it right to say that the ball has gone out of the train's frame to the platform's frame of reference?*

T9 You are making the same error as in S7. As I have already emphasized, a train's frame is not localized or bounded by the train. There is no meaning to the size, territory or boundary of a frame of reference. Every phenomenon can be described relative to any frame of our choice. It is wrong to think that a certain phenomenon 'belongs' to a particular frame. A physical phenomenon stands by itself. Relative to different frames, we have different descriptions of it. Thus the throwing of a ball from a running train to the platform is a phenomenon that can be described relative to both the train's frame and the platform's frame.

As another example, suppose on board of ship at sea, a child is playing with a ring. It is again incorrect to say that this phenomenon 'takes place' in the ship's frame of reference. The phenomenon is apart from the frames. It can be described relative to the ship's frame or the harbour's frame or the frame of reference of a flying aeroplane or any other frame.

In short, we have material bodies in the universe in all kinds of states of motion. With each body there is an associated frame of reference which shares only the motion, but no other characteristic, of the body. Thus we can associate a frame of reference with the sun, the earth, a speeding rocket, an electron in an atom etc., and every phenomenon in the universe can be described relative to any of these frames.

Transformations between frames

S10 *I am now clear about the notion of frames of reference. But suppose I know the description of a phenomenon in one frame, and wish to know what it will be in another frame, what should I do?*

T10 For that you will need to know how physical quantities like position, velocity, acceleration, etc. transform from one frame to another. That, naturally, depends on what frame you are transforming to. Let us consider various cases:

I. Translation in space and time

This involves a mere shift of origin as shown in Fig. 5.2; there is no relative motion between S and S'. Clearly, the position vector $\boldsymbol{r}'$ of a point relative to S' is related to the position vector $\boldsymbol{r}$ of the same point relative to S by the relation

$$\boldsymbol{r} = \boldsymbol{r}' + \boldsymbol{R}, \tag{1}$$

where $\boldsymbol{R}$ is the position vector of O' relative to O. In component-form this equation reads:†

$$r_i = r_i' + R_i, \qquad i = 1, 2, 3. \tag{2}$$

We can also include a shift in the zero of the clocks used in S and S'; i.e. the same instant has time co-ordinates t and t', which differ by a constant.

$$t = t' + T. \tag{3}$$

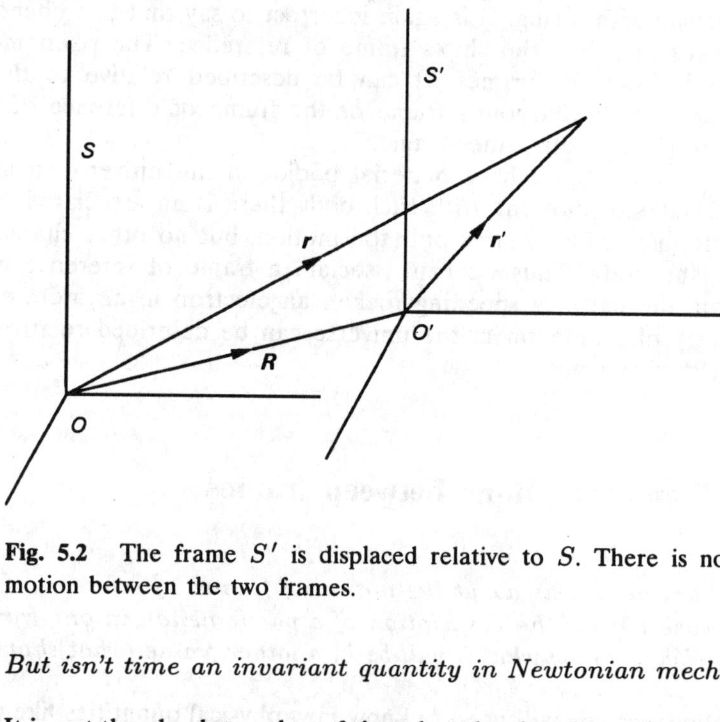

Fig. 5.2 The frame S' is displaced relative to S. There is no relative motion between the two frames.

S11 *But isn't time an invariant quantity in Newtonian mechanics?*

T11 It is not the absolute value of time, but the time interval between two given events that is invariant in Newtonian theory. Two observers can have different zero settings of their clocks, and therefore can assign different values of times to a given event. For instance, the <u>same</u> instant of time is 5 a.m. in England and 10.30 a.m. in India. However, the duration of say one hour is the same at both places. This also follows from the above equations, since T is fixed:

$$t_2 - t_1 = t_2' - t_1'. \tag{4}$$

† We adopt the standard convention that components 1, 2 and 3 refer to the x, y and z components, respectively. Thus, $r_2 = y$, $r_1' = x'$ etc.

Now, the definitions of velocity and acceleration involve time intervals which, as we have seen, are invariant. Thus, since $\boldsymbol{R}$ is a fixed vector, Eqs. (1) and (4) give

$$\boldsymbol{v} = \boldsymbol{v}' \quad \text{and} \quad \boldsymbol{a} = \boldsymbol{a}'$$

which, in terms of components, read as

$$v_i = v_i'; \quad a_i = a_i' \quad i = 1,\ 2,\ 3.$$

S12 *You have considered the axes of S′ parallel to S. What if the axes are oriented differently?*

T12 You have anticipated the next case that I am going to consider.

II. Translation and rotation

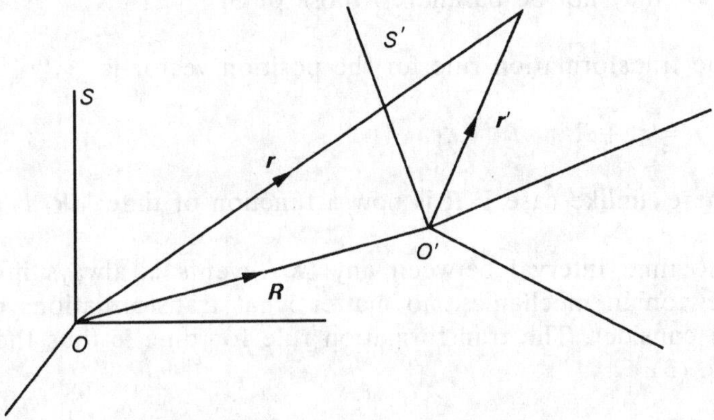

Fig. 5.3 The frame S' is displaced and rotated relative to S. The two frames have no relative motion.

In this case, the vector relation [Eq. (l)] remains the same (See Fig. 5.3). The component-form [Eq. (2)], however, modifies to

$$\boldsymbol{r}_i = \sum_{j=1}^{3} U_{ij}\boldsymbol{r}_j' + \boldsymbol{R}_i, \qquad i = 1,\ 2,\ 3$$

where U_{ij} is the transformation matrix for rotation that we discussed in Chapter 2. Likewise, the vector relations for velocity and acceleration are the same as in Case 1, but in terms of components they are

$$v_i = \sum_{j=1}^{3} U_{ij}v_j'; \qquad a_i = \sum_{j=1}^{3} U_{ij}a_j'; \qquad i = 1,\ 2,\ 3.$$

For pure rotations i. e., rotations without any translation, simply put $R_i = 0$ in the transformation relation for co-ordinates above.

Remember, as in case I, there is no relative motion between S and S' here.

S13 *What happens if there is a relative motion between S and S'?*

T13 This indeed is the situation that most interests us in this chapter. We shall consider two distinct types of relative motion: relative translational motion and relative rotational motion.

III. Relative translational motion

Here, S' has pure translational motion relative to S, i.e. O' moves with a velocity $\boldsymbol{V}$ with respect to O, but the orientation of the axes of S' remains *fixed* relative to S (See Fig. 5.4). Of course, the axes of S' may not be parallel to those of S.

The transformation rule for the position vector is

$$\boldsymbol{r} = \boldsymbol{r}' + \boldsymbol{R} \tag{5}$$

where, unlike case I, $\boldsymbol{R}$ is now a function of time: $d\boldsymbol{R}/dt = \boldsymbol{V}$.

The time interval between any two events is always invariant in Newtonian mechanics, no matter what transformations of frames we consider. The transformation rule for time is thus the same as Eq. (3):

$$t = t' + T.$$

From Eqs. (5) and (3) we get the transformation rules for velocity and acceleration:

$$\boldsymbol{v} = \boldsymbol{v}' + \boldsymbol{V} \quad \text{and} \quad \boldsymbol{a} = \boldsymbol{a}' + \boldsymbol{\alpha},$$

where $\boldsymbol{\alpha} = d\boldsymbol{V}/dt$ stands for the acceleration of O' with respect to O. In component form,

$$\boldsymbol{v}_i = \sum_{j=1}^{3} \boldsymbol{U}_{ij}\boldsymbol{v}_j' + V_i \quad \text{and} \quad \boldsymbol{a}_i = \sum_{j=1}^{3} \boldsymbol{U}_{ij}\boldsymbol{a}_j' + \boldsymbol{\alpha}_i \quad i = 1, 2, 3.$$

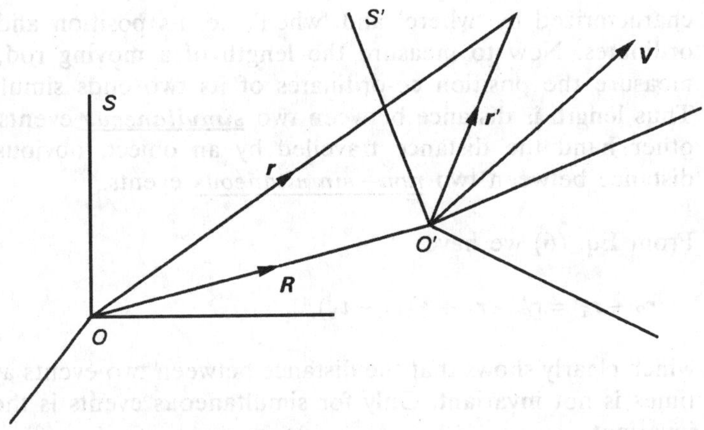

Fig. 5.4 The frame S' moves relative to the frame S with velocity V. The orientation of the axes of S', however, remains fixed throughout.

The simpler case of *uniform relative motion* between S and S' is especially important. Here $\boldsymbol{V}$ is constant so that $\boldsymbol{\alpha}$ is zero. The transformation rules for various quantities, in this case become:

$$\boldsymbol{r} = \boldsymbol{r}' + \boldsymbol{R}_0 + \boldsymbol{V}t; \qquad t = t' + T$$

$$\boldsymbol{v} = \boldsymbol{v}' + \boldsymbol{V}; \qquad \boldsymbol{a} = \boldsymbol{a}' \tag{6}$$

where $\boldsymbol{R}_0$ gives the location of O' relative to O at $t = 0$. Equations (6) are sometimes known as *Galilean transformations*.

For simplicity, one often takes $T = 0$. Further, for convenience, we take O and O' to be coincident at $t = t' = 0$. The transformation equations for position vector and time then reduce to

$$\boldsymbol{r} = \boldsymbol{r}' + \boldsymbol{V}t; \qquad t = t'. \tag{7}$$

S14 *I have a query. You said that the time interval between any two events is invariant in Newtonian theory. Isn't the length or distance also an invariant?*

T14 You must be careful here. The length of an object say a rod, is an invariant. The distance between two events, on the other hand, is *not* an invariant in general.

S15 *I don't quite get you.*

T15 This seems to be a source of confusion among students. It will help if you always keep in mind one obvious point. An event is characterized by 'where' and 'when', i.e. its position and time co-ordinates. Now to measure the length of a moving rod, we must measure the position co-ordinates of its two ends simultaneously. Thus length is distance between two *simultaneous* events.† On the other hand the distance travelled by an object, obviously, is the distance between two *non–simultaneous* events.

From Eq. (6) we have

$$\boldsymbol{r}_2 - \boldsymbol{r}_1 = \boldsymbol{r}_2' - \boldsymbol{r}_1' + \boldsymbol{V}(t_2 - t_1), \tag{8}$$

which clearly shows that the distance between two events at different times is not invariant. Only for simultaneous events is the distance invariant:

$$\boldsymbol{r}_2 - \boldsymbol{r}_1 = \boldsymbol{r}_2' - \boldsymbol{r}_1' \quad \text{when} \quad t_2 = t_1.$$

Thus the length of an object is an invariant in Newtonian relativity —something that everyone knows intuitively.

S16 *This appears to be an important point. Can you illustrate it?*

T16 Consider a tube of length 2 m in a moving train as shown in Fig. 5.5. How much do you think is the length of the tube for a ground observer?

S17 *Obviously 2 m; the length is an invariant quantity.*

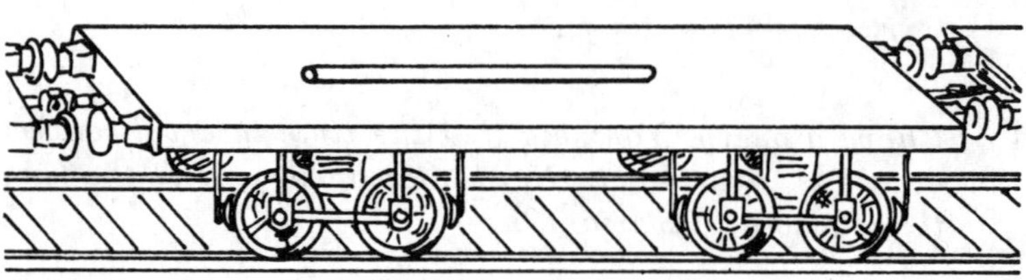

Fig. 5.5 A tube on a moving railway carriage.

† To find the length of a stationary rod, the measurements need not be simultaneous. But we are talking of transformation of length from one frame to another. A rod at rest relative to one frame will be moving relative to another frame. So, in general, we must consider simultaneous measurements for determining length.

T17 That is right. Now suppose I roll a ball down from one end of the tube to the other. How much do you think is the distance travelled by the ball relative to the train observer, and the ground observer?

S18 *Obviously the same, 2m for both, because the ball has rolled down the tube and the length of the tube is 2 m for both.*

T18 Your answer is wrong. This is because the event of the ball starting at one end of the tube and the event of its reaching the other end are not simultaneous. The distance between these non-simultaneous events is not the same for the two observers [See Eq. (8)]. It is equal to 2 m for the train observer, but different from 2 m for the ground observer. (For example, if the ball moves in the same direction as the train, the distance travelled by the ball for the ground observer is greater than 2 m.)

S19 *How about the time the ball takes to travel from one end of the tube to the other? Will it be different for the two observers?*

T19 Of course not. Do not forget that the time interval between any two events is absolute in Newtonian relativity; it is the same for every observer.

S20 *Then how is it that the distance travelled by the ball relative to the two observers is different?*

T20 Simple. Because the speed of the ball is different for them. Let us see this more explicitly. Suppose the ball moves at a speed of 1 m/s relative to the train. Relative to the train observer, clearly, the distance travelled is 2 m and the time of travel is 2 s.

Now suppose the train moves at a speed of 10 m/s. Obviously, the ball moves at a speed of 11 m/s relative to the ground. Since time intervals are invariant, the time of travel relative to the ground observer is the same as that for the train observer, i.e. 2 s. Clearly, therefore, the distance travelled by the ball for the ground observer is 11 m/s $\times$ 2 s = 22 m. (If the ball had moved in the direction opposite to the train's motion, the answer would be 9 m/s $\times$ 2 s = 18 m.)

S21 *I get the point. Any other example?*

T21 Consider a river flowing with uniform velocity. Suppose there are two trees on the opposite banks of the river such that the line joining them is perpendicular to the flow.

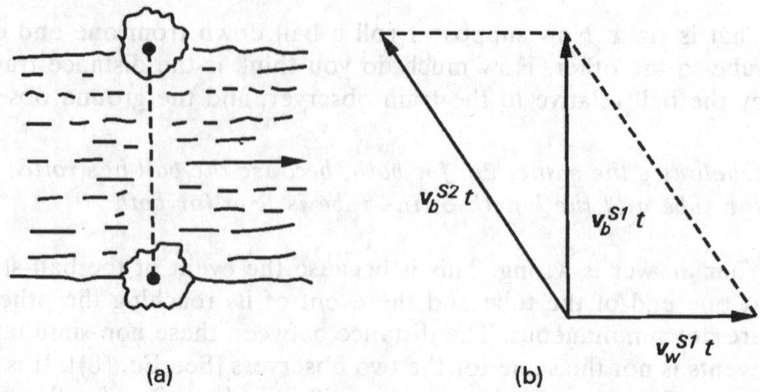

Fig. 5.6 The displacement of the boat relative to water ($v_b^{S2}t$) and that of water relative to the ground ($v_w^{S1}t$) add up to the displacement of the boat relative to the ground ($v_b^{S1}t$) in accordance with Eq. (6).

See Fig. 5.6(a). A boat travels along this line from one tree to the other. Now consider two frames of reference: the ground frame ($S1$) and a frame moving with the river ($S2$). Is the distance between the two trees the same for both the observers?

S22 *No, because distance is not invariant as you have just said above.*

T22 You are wrong. Note that the trees have no relative motion. So, the distance between them simply means the length of a rod connecting them, or in this example, the width of the river. This, as you know, is invariant.

Now consider the distance travelled by the boat. Is it the same for the two frames S_1 and S_2?

S23 *I guess so. The boat travels the width of the river, which is the same for both S_1 and S_2.*

T23 You are wrong again. The distance travelled by the boat is the distance between two <u>*non–simultaneous events*</u>. This distance is different for the two frames in relative motion; the reason being the same as that in the previous example. The time of travel for the boat is the same for S_1 and S_2. The velocity of the boat, however, is different relative to the two frames and, hence, so are the distances. See Fig. 5.6(b).

S24 *I think it is clear now. Let us return to the general discussion about transformation of quantities between frames. We have so far taken the orientation of axes of S to be S′ as fixed. What happens if the relative orientation of the axes changes in time?*

T24 This is an important case that we consider next.

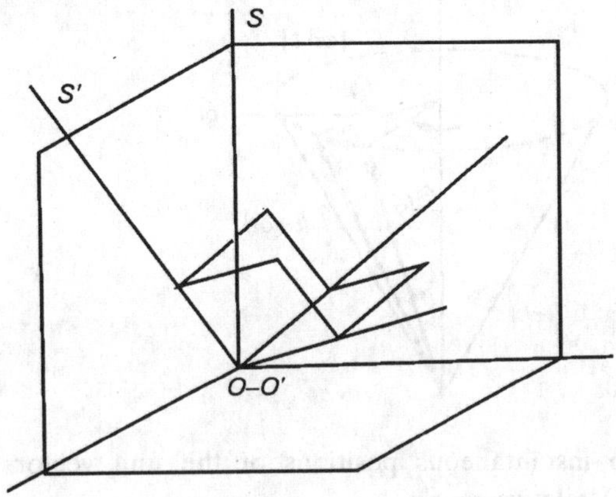

Fig. 5.7 The frame S' rotates freely with its orign O' coincident with the origin O of S at all times.

IV. Relative rotational motion

Here, the axes of S' rotate relative to S with an angular velocity $\boldsymbol{\omega}$, while the origins of S and S' remain coincident (See Fig. 5.7). Evidently, the position vector of any point, $\boldsymbol{r}$, has different components relative to the two frames. Moreover, the components vary at different rates because of the relative motion between the frames. The relations between velocities and accelerations, as measured relative to the two frames, are derived in Box 5.1. This derivation is a little involved and you may skip it in the first reading. But the results are important for our subsequent discussion. They are as follows:

The relation between the velocity $\boldsymbol{v} = (d\boldsymbol{r}/dt)_s$ relative to S and the velocity $\boldsymbol{v}' = (d\boldsymbol{r}/dt)_{s'}$, relative to S' is

$$\boldsymbol{v} = \boldsymbol{v}' + \boldsymbol{\omega} \times \boldsymbol{r} \tag{9}$$

The acceleration $\boldsymbol{a} = (d\boldsymbol{v}/dt)_s$ relative to S and the acceleration $\boldsymbol{a}' = (d\boldsymbol{v}'/dt)_s$ relative to S' are related by the equation

$$\boldsymbol{a} = \boldsymbol{a}' + 2\boldsymbol{\omega} \times \boldsymbol{\nu}' + \boldsymbol{\omega} \times (\boldsymbol{\omega} \times \boldsymbol{r}) + \frac{d\boldsymbol{\omega}}{dt} \times \boldsymbol{r} \tag{10}$$

Box 5.1

A vector which is constant relative to S', obviously varies with time relative to S. Consider the unit vector $\hat{i}'$ along the x'-axis of the S' frame, which

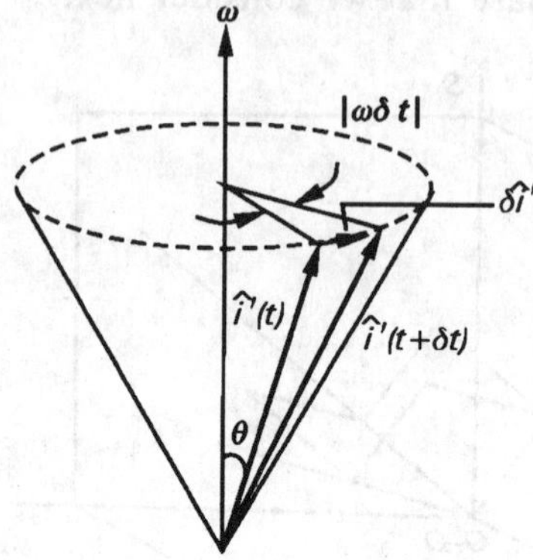

Fig. 5.8 Two instantaneous positions of the unit vector $\hat{i}'$ (which is constant in S') relative to S.

rotates with the frame with an angular velocity $\boldsymbol{\omega}$. Two instantaneous positions of $\hat{i}'$ relative to S, viz., $\hat{i}'(t)$ and $\hat{i}'(t+\delta t)$, are shown in Fig. 5.8. It is clear from the figure that

$$|\delta\hat{i}'| = \sin\theta |\boldsymbol{\omega}\delta t|$$

and hence,

$$\left|\frac{d\hat{i}'}{dt}\right| = |\boldsymbol{\omega} \times \hat{i}'|.$$

From the figure, the direction of $d\hat{i}'/dt$ matches that of $\boldsymbol{\omega} \times \hat{i}'$. We can thus write

$$\left(\frac{d\hat{i}'}{dt}\right)_s = \boldsymbol{\omega} \times \hat{i}'. \qquad (11a)$$

Similarly,

$$\left(\frac{d\hat{j}'}{dt}\right)_s = \boldsymbol{\omega} \times \hat{j}' \quad \text{and} \quad \left(\frac{d\hat{k}'}{dt}\right)_s = \boldsymbol{\omega} \times \hat{k}'. \qquad (11b)$$

The position vector $\boldsymbol{r}$ of any point has different components in S and S' since the two frames differ in the basis vectors:

$$\begin{aligned}\boldsymbol{r} &= x\hat{i} + y\hat{j} + z\hat{k} \\ &= x'\hat{i}' + y'\hat{j}' + z'\hat{k}'.\end{aligned}$$

The rate of change of $\boldsymbol{r}$ relative to S can be calculated using any one of the expansions. If we choose the first one, $\hat{i}$, $\hat{j}$ and $\hat{k}$ are constant vectors, and we get the familiar result:

$$\left(\frac{d\boldsymbol{r}}{dt}\right)_s = \frac{dx}{dt}\hat{i} + \frac{dy}{dt}\hat{j} + \frac{dz}{dt}\hat{k}.$$

If we choose the second expansion, the fact that $\hat{i}'$, $\hat{j}'$ and $\hat{k}'$ are not constant vectors relative to S must be taken into account. Hence,

$$\begin{aligned}\left(\frac{d\boldsymbol{r}}{dt}\right)_s &= \frac{dx'}{dt}\hat{i}' + \frac{dy'}{dt}\hat{j}' + \frac{dz'}{dz}\hat{k}' \\ &\quad + x'\left(\frac{d\hat{i}'}{dt}\right)_s + y'\left(\frac{d\hat{j}'}{dt}\right)_s + z'\left(\frac{d\hat{k}'}{dt}\right)_s.\end{aligned} \tag{12}$$

Since $\hat{i}'$, $\hat{j}'$ and $\hat{k}'$ are constant relative to S', clearly,

$$\left(\frac{d\boldsymbol{r}}{dt}\right)_{s'} = \frac{dx'}{dt}\hat{i}' + \frac{dy'}{dt}\hat{j}' + \frac{dz'}{dt}\hat{k}'.$$

Equations (11) and (12) then give

$$\left(\frac{d\boldsymbol{r}}{dt}\right)_s = \left(\frac{d\boldsymbol{r}}{dt}\right)_{s'} + \boldsymbol{\omega} \times \boldsymbol{r},$$

which is Eq. (9). Indeed a similar relation is true for the rate of change of any vector $\boldsymbol{A}$ relative to the two frames:

$$\left(\frac{d\boldsymbol{A}}{dt}\right)_s = \left(\frac{d\boldsymbol{A}}{dt}\right)_{s'} + \boldsymbol{\omega} \times \boldsymbol{A} \tag{13}$$

In order to find the relation between accelerations relative to the two frames, we differentiate Eq. (9) relative to S:

$$\left(\frac{d\boldsymbol{v}}{dt}\right)_s = \left(\frac{d\boldsymbol{v}'}{dt}\right)_s + \left(\frac{d}{dt}\boldsymbol{w} \times \boldsymbol{r}\right)_s. \tag{14}$$

For the first term on the right-hand side we can use Eq. (13), with A replaced by $\boldsymbol{v}'$:

$$\left(\frac{d\boldsymbol{v}'}{dt}\right)_s = \left(\frac{d\boldsymbol{v}'}{dt}\right)_{s'} + \boldsymbol{\omega}\times\boldsymbol{v}'. \tag{15}$$

The second term on the right-hand side of Eq. (14) can be evaluated in a straightforward manner:

$$\begin{aligned}\left(\frac{d}{dt}\boldsymbol{\omega}\times\boldsymbol{r}\right)_s &= \left(\frac{d\boldsymbol{\omega}}{dt}\right)_s \times\boldsymbol{r} + \boldsymbol{\omega}\times\left(\frac{d\boldsymbol{r}}{dt}\right)_s \\ &= \left(\frac{d\boldsymbol{\omega}}{dt}\right)_s \times\boldsymbol{r} + \boldsymbol{\omega}\times\boldsymbol{v}' + \boldsymbol{\omega}\times(\boldsymbol{\omega}\times\boldsymbol{r}).\end{aligned} \tag{16}$$

We have used Eq. (9) while writing the second step. Using Eqs. (15) and (16) in Eq. (14) we obtain

$$\boldsymbol{a} = \boldsymbol{a}' + 2\boldsymbol{\omega}\times\boldsymbol{v}' + \boldsymbol{\omega}\times(\boldsymbol{\omega}\times r) + \frac{d\boldsymbol{\omega}}{dt}\times\boldsymbol{r},$$

which is Eq. (10).

Inertial and non-inertial frames/pseudo-forces

S25 *Now I understand how to make transformations from one frame to another. Let me turn to another theme. Frames of reference are classified as inertial or non-inertial. How do we decide which frame is inertial and which non-inertial?*

T25 The prescription is simple, and it was explained in Chapter 4. If, relative to a given frame of reference, the I law is valid then the frame is inertial. If it is difficult to have a no-force situation, we can test the inertial character of a frame by using the second law $\boldsymbol{F} = m\boldsymbol{a}$, where $\boldsymbol{F}$ is the resultant of all the material forces acting on the test body. If the equation is satisfied, the frame is inertial; otherwise, it is non-inertial.

S26 *Then what is the law of motion in a non-inertial frame?*

T26 In a non-inertial frame, some additional terms need to be put in so that we get a sort of modified form of the II law

$$m\boldsymbol{a} = \boldsymbol{F} + \boldsymbol{F}_p. \tag{17}$$

$\boldsymbol{F}_p$ are called pseudo-forces which are to be distinguished from material forces such as gravitational, electromagnetic, contact forces etc. They are still qualified as 'forces' because they sit beside the material forces $\boldsymbol{F}$ in the modified II law, and together the two are responsible for the observed acceleration.

S27 *I am not clear about the nature and origin of pseudo-forces in a non-inertial frame. Please clarify.*

T27 Psudo-forces are frame-dependent forces. This means the following:

In a given non-inertial frame, if you experiment on different bodies, it is found that, in order to account for the observed acceleration of the bodies, besides the material forces $\boldsymbol{F}$, you have to introduce certain additional terms like $\boldsymbol{F}_p$. These terms not only depend on the mass, velocity and acceleration of a body, they also involve additional parameters that are common to all bodies. Now what can these parameters possibly refer to? We have already said that all the material forces are included in $\boldsymbol{F}$. So the parameters in $\boldsymbol{F}_p$ which are common to all bodies cannot possibly refer to any material agencies. They must refer to the properties of the frame of reference itself.

S28 *I did not quite get you. Can you be more specific?*

T28 You will understand the origin of the pseudo-forces better if you start with the II law of motion in an inertial frame S, and see how it looks like in another frame moving relative to S. This is quite straightforward to do, since you already know how kinematical quantities transform from one frame to another.

1) First consider a frame S' that is moving with uniform velocity $\boldsymbol{V}$ with respect to S. The law of motion in S is:

$$m\boldsymbol{a} = \boldsymbol{F}. \tag{18}$$

The acceleration $\boldsymbol{a}'$ of a body relative to S' and the acceleration $\boldsymbol{a}$ of the same body relative to S are equal when the relative motion between S and S' is uniform. In addition, you must realize that the force $\boldsymbol{F}$ remains unchanged as we go from S to S'. This is because material forces in mechanics depend only on the (instantaneous) <u>relative</u> positions and velocities of bodies; and it can be seen from Eq. (6) that these quantities are invariant under the transformations under consideration. Further, the mass of a body is an invariant quantity in Newtonian mechanics. Consequently, the law of motion in S' is:

$$m\boldsymbol{a}' = \boldsymbol{F},$$

which is identical in form as in S.

This is a very significant thing. It shows that if S is an inertial frame, any frame moving with uniform velocity relative to S is also inertial; i.e. the form of the II law is the same in all inertial frames.

2) Now consider a frame S' that has arbitrary translational motion relative to S. As seen in T13 the relation between $\boldsymbol{a}$ and $\boldsymbol{a}'$ is

$$\boldsymbol{a} = \boldsymbol{a}' + \boldsymbol{\alpha}$$

where, $\boldsymbol{a}$ is the acceleration of S' relative to S. Using this relation in the law of motion in S [Eq. (18)] we get

$$m\boldsymbol{a}' = \boldsymbol{F} - m\boldsymbol{\alpha}. \tag{19}$$

So you see, in contrast to the earlier case, the form of the II law in S' is different from that in S.

S' is, thus, not an inertial frame. The additional term, $-m\boldsymbol{\alpha}$, appearing in the law of motion is what is called a pseudo-force. The parameter $\boldsymbol{\alpha}$ characterizes the non-inertial frame S'. The pseudo-force $-m\boldsymbol{\alpha}$ is thus a frame-dependent force.

3) Lastly, consider a frame S' rotating with respect to S with an angular velocity $\boldsymbol{\omega}$. The relation between accelerations relative to the two frames is given by Eq. (10). Using this in Eq. (18) we obtain

$$m\boldsymbol{a}' = \boldsymbol{F} - 2m\boldsymbol{\omega} \times \boldsymbol{v}' - m\boldsymbol{\omega} \times (\boldsymbol{\omega} \times \boldsymbol{r}) - m\frac{d\boldsymbol{\omega}}{dt} \times \boldsymbol{r}. \tag{20}$$

Once again we see that the form of the II law is not the same as in the inertial frame S.

Thus the rotating frame S' is not an inertial frame. The additional terms in its law of motion involve m, $\boldsymbol{v}$, $\boldsymbol{r}$ of the body. But they also involve a parameter $\boldsymbol{\omega}$ which does not depend on the body; it characterizes the frame itself. They are the frame-dependent forces mentioned in T27. The second term on the right hand side of Eq. (20) is called the Coriolis force. The third term is the centrifugal force, which is independent of velocity. The last term involves the rate of change of $\boldsymbol{\omega}$. In most applications of interest, $\boldsymbol{\omega}$ is either constant or changes very slowly. We shall ignore the last term in our subsequent discussion.

To recapitulate, in an inertial frame $m\boldsymbol{a} = \boldsymbol{F}$. In a non-inertial frame, $m\boldsymbol{a} = \boldsymbol{F} + \boldsymbol{F}_p$, where $\boldsymbol{F}_p$ contains frame-dependent parameters such as $\boldsymbol{\alpha}$ and $\boldsymbol{\omega}$.

S29 *I have a question here. Let us return to the frames S and S' considered above in Case 2. Now suppose I consider a third frame S'' which has the same acceleration $\boldsymbol{\alpha}$ relative to S as S' has, but may have a different velocity. Is it correct to say that S'' is non-inertial with respect to S, but is inertial with respect to S', since the relative acceleration between S' and S'' is zero?*

T29 Your question clearly shows that you haven't got the essential point of the preceding discussion. I suggest you go through it again. In particular, do read T25 to T27.

The thing that you must understand is that the inertial or non-inertial character of a frame of reference is a property of the frame itself, which tells you whether or not measurements relative to it obey the laws of motion in their simple form. It is possible to establish this property of a frame empirically without reference to any other frame, as I have already explained in T27.

However, as we saw, if a frame S is given to be inertial, any frame moving uniformly relative to S is also inertial; and a frame accelerating or rotating relative to S is non-inertial. This by no means implies that inertial or non-inertial character is a bilateral property of the two frames. The statement that a frame is inertial or non-inertial with respect to some other frame is absurd.

Thus, in your example, both S' and S'' are non-inertial, since they both accelerate relative to S which is given to be inertial.

S30 *That is quite clear now. But what is so significant about all this? I mean what is there to make so much fuss about?*

T30 The significant thing is that nature admits of a special class of frames in which the laws of motion are especially simple. In other frames the laws get complicated by frame-dependent forces. The famous Newton's pail experiment is a vivid demonstration of just this fact (See Box 5.2). In short, *inertial frames are the privileged frames of nature.*

Box 5.2

Newton's Pail Experiment

A pail (bucket) of water is suspended from the ceiling by a rope. The rope is twisted about itself to the maximum extent possible and is released. The pail begins to rotate, accelerates to its maximum angular velocity ω (when the rope is fully untwisted), and subsequently deaccelerates to rest as the rope gets twisted in the opposite sense.

Consider two observers S and S' observing the surface of the water in the pail. S is at rest in the laboratory. The frame of reference of S' rotates with a constant angular velocity ω (the same as the maximum angular velocity of the pail). Let us see what laws the two observers report on the shape of the water surface in the pail.

Relative to S:

The surface of the water in a stationary pail is a plane; the surface of water in a rotating pail is curved (paraboloid), the curvature increasing with the angular speed of the pail.

Relative to S':

The surface of the water in a stationary pail is a paraboloid with a particular value for the curvature; the surface of water in a pail is a plane when it has a particular angular velocity ($-\omega$).

This experiment clearly shows the inequivalence of the frames of reference S and S'. The laws of physics are evidently simpler in S than in S'. S is an inertial frame and S' is non-inertial.

S31 *Is that not against the principle of relativity which says that no frame is more privileged than the other?*

T31 That sweeping statement is neither true in Newtonian mechanics nor in Einstein's special relativity.† What is true in both is that all *inertial* frames are equivalent. Non-inertial frames are not equivalent to inertial frames from the viewpoint of the laws of physics as we have seen in T28 .

S32 *What precisely is meant by the statement that two frames are equivalent?*

† It is true in Einstein's general relativity, in a sense which cannot be explained here.

T32 Two frames are said to be equivalent if the laws of physics have the same form in both. Beware, this does not mean that all the physical quantities have the same values in both. For example, for two inertial frames in relative motion, the position vector $\boldsymbol{r}$ and velocity $\boldsymbol{v}$ of a body relative to S will not be the same as $\boldsymbol{r}'$ and $\boldsymbol{v}'$ of the same body relative to S'. In general, in equivalent frames, it is the relations between the quantities which are the same, not the values of the quantities themselves.

The phrase 'no inertial frame is more privileged than others' is just a rewording of the statement that 'all inertial frames are equivalent'.

S33 *But in actual practice, we do prefer frames. For example, if I have to describe the motion of a ball thrown up by a child in a uniformly moving train, it is better to use the train's frame than the ground's frame. In the former, the trajectory of the ball is a straight line, while in the latter it is a parabola. Isn't the train's frame more privileged than the ground's frame in this case?*

T33 This is a very natural pitfall. Your error lies in equating a convenient frame for a given phenomenon to a privileged frame from the point of view of the laws of physics. We have already stressed that equivalence of frames does not mean equality of measured values of quantities. In other words, the description of a phenomenon may be different in two frames, even though the frames are equivalent. And one description could well be simpler than another.

To put it more rigorously, the trajectory of a particle is determined not only by the laws of motion but also by its initial conditions. The initial conditions for a particular problem may be simpler in one frame than in another even if the laws are identical. In your example, the horizontal velocity of the ball has a simple value (namely zero!) in the train's frame but is non-zero in the ground's frame. That is why, even though the laws of motion are identical on the ground and in the uniformly moving train, the trajectory of the ball is different.

In short, remember not to confuse equivalence of frames with the identity of descriptions of physical phenomena.

Uniform motion is relative; acceleration/rotation are absolute.

S34 *There is a standard phrase: 'uniform motion is relative' that one often comes across so often. Can you clarify just what this phrase means?*

T34 The basic relevant facts are already known to you. Now physicists are not always dry people. They like to put an important idea across in different catchy ways, partly to convey the essence of the idea, and partly with a view to communicate technical points in non-technical language.

Let us see what one means by what has become the standard cliche of relativity: uniform motion is relative.

Suppose I am in an inertial frame S and you are in a frame S' moving uniformly relative to me. Now suppose I am both ignorant and arrogant, and claim that I am 'really' at rest and it is you who is really moving. I claim that my motion that you observe is apparent, not real. You, on the other hand, are also adamant. You insist that it is you who is really at rest, and that I am actually moving. How do we solve this deadlock? We approach a physicist.

The physicist's reply will be something like this:

"Each of you can think himself to be at rest and the other to be moving, if that satisfies your petty egos. But physics-wise I have no way of deciding in anyone's favour. Relative to S, S' is moving. Relative to S', S is moving in the opposite direction. Both statements are correct. Since the laws of physics are identical in S and S', there is no way in physics to test whether one of them is 'actually' at rest. The notion of absolute rest, therefore, is inadmissible in physics".

After explaining thus, the physicist summarizes the para with a catchy phrase 'uniform motion is relative', and calls it the fundamental principle of relativity.

S35 *But still I can't help feeling that some motions are real and some apparent. For example, in a train's frame, the motion of the ground and everything else does seem to be apparent. On the other hand, the motion of the train relative to the ground seems real. Why does that feeling arise even when I know the principle of relativity?*

T35 That feeling is perhaps the result of a number of reasons. First, the earth's frame is our customary frame of reference. Further, the earth is in free fall under the sun's gravity; we do not need to expend any energy to keep it moving. On the other hand, we are aware that to have a train start from rest and to keep it in uniform motion (against friction) we need to expend energy. There is also an asymmetry in the situation with regard to size; the train is much smaller than the surroundings, and it is more natural to concede 'reality' to the motion of the train than to that of the surroundings. Another reason is that initially the train needs to be accelerated, and its subsequent motion is rarely, if ever, strictly uniform.

In this connection let me bring a common experience to your notice. From a train that is just beginning to move, when its motion is smooth, if you look out of a window at another train on an adjacent track, it is difficult to judge if it is your train that is moving or the other. This happens because the various reasons suggested in the para above are not relevant to this experience.

But whatever the cognitive reasons for the feeling regarding the real and apparent motion, be clear about one thing. As long as the train is in uniform motion, there is no way in physics to say that the motion of the train relative to the ground is real and that of the ground and surroundings relative to the train is apparent.

S36 *Why do you insist on the word 'uniform' in your statement of the principle of relativity? Why can't we say that motion, in general, is relative?*

T36 You already know the answer to that if you care to go back to T28. We know that when a frame S' accelerates or rotates with respect to an inertial frame S, S' is not inertial. The laws in S and S' are then no longer identical. Unlike uniform motion, acceleration and rotation are absolute concepts.

S37 *I didn't get you.*

T37 Go back to our quarrel in T34 with a slight modification. I continue to be in the inertial frame S but now your frame S' is accelerating relative to me. I claim that I am unaccelerated and that it is you who is actually accelerating. You, adamant as before, insist that you are unaccelerated and that I am actually accelerating in the opposite direction. This time when we approach a physicist, he will be on my side.

S38 *Why? Isn't it true that if relative to S, S' has an acceleration a, then relative to S', the acceleration of S is $-a$?*

T38 That is quite correct kinematically. There is no quarrel on what each of us observes about the other. The issue is whether there is a way in physics to judge if a frame of reference is <u>*actually*</u> accelerated or not.

S39 *How will the physicist decide this issue?*

T39 The physicist's reply will be something like this:

"This time, unlike the earlier quarrel, I have a way of settling the issue. The law of motion is different in S and S'. The II law in S' [Eq. (19)] involves a frame-dependent parameter $\boldsymbol{\alpha}$ that stands for the acceleration of S' relative to any inertial frame. In S the law of motion has no such frame-dependent parameter. So an inertial frame like S should be considered unaccelerated. Acceleration of a frame is thus an absolute concept, i.e. in physics there is a way of deciding that a particular frame of reference is or is not *actually* accelerated.

An exactly similar argument goes through if a frame S' is rotating relative to an inertial frame S. The II law in S' [Eq. (20)) involves a frame-dependent parameter $\boldsymbol{\omega}$ which stands for the angular velocity of the frame relative to any inertial frame. In S the law of motion has no such frame-dependent parameter. So any inertial frame of reference like S should be considered non-rotating, and it is S' that is really rotating with angular velocity $\boldsymbol{\omega}$. Rotation is thus an absolute concept."

S40 *Can you recapitulate the various points in the discussion so far?*

T40 If a frame is inertial, so is any other frame moving uniformly relative to it. Inertial frames are equivalent to one another, i.e. they have the same form of the II law. This prevents us from singling out, among these, any one frame which can be claimed to be really at rest.

But inertial frames are not equivalent to non-inertial frames. The presence of frame dependent forces (pseudo-forces) in the II law for non-inertial frames allows one to claim that these are actually accelerating or rotating, while the inertial frames are unaccelerated and non-rotating.

Perhaps the best way of saying all this is that, by experiments described relative to a given frame of reference, it is possible to determine the intrinsic acceleration/rotation parameters of the frame, but not its velocity. If the acceleration/rotation parameters are found to be zero, the frame is inertial, or else non-inertial.

S41 *I have another query. Each day we see the sun moving across the sky. So, relative to the earth the sun goes around the earth, while relative to the sun the earth rotates about its axis. Why, then, do we insist that the rotation of the sun is apparent, and it is indeed the earth that rotates?*

T41 This is just an application of the ideas already discussed. Rotation, unlike uniform motion, is not relative; it is an absolute idea. That is, there is a way to deduce whether it is the sun or the earth that is actually rotating. Take the sun's frame of reference and consider the motion of the earth and other planets. This can be described by the II law of Newton (combined with the law of gravitation) with no (or negligible) addition of pseudo-force terms. Thus the sun's frame is inertial. Now, take the earth's frame and consider the motion of the sun and other planets. One finds that one needs to include in the II law of motion pseudo-force terms which contain an angular velocity parameter. This parameter is the same whether we consider the motion of the sun or some planet; it must be interpreted as the angular velocity of the frame itself, i.e. that of the earth. Thus the earth's frame is non-inertial; it is the earth that actually rotates, not the sun.†

S42 *But we do consider the earth's frame as inertial for most terrestrial experiments. How can the same frame of reference be inertial for one purpose and non-inertial for another purpose?*

T42 It is a question of the relative magnitudes of the pseudo-force terms and the material force terms which appear in the II law of motion for a non-inertial frame. You see, the ratio F_p/F depends not only on the parameters of the frame such as $\boldsymbol{\alpha}$, $\boldsymbol{\omega}$ etc., but also on other things like mass, position and velocity of the body and other characteristics of material forces.

For astronomical observations relative to the earth's frame, the pseudo-forces cannot be neglected. Indeed, they mainly account for the observed daily rotation of the stars. For most (but not all) terrestrial experiments, however, the ratio F_p/F is negligible. For these situations, therefore, the earth's frame can be considered an approximately inertial frame.

In short, inertial or non-inertial character of a frame is a matter of degree, and depends on the realm of your experiments. An ideal inertial frame, for which all the acceleration/rotation parameters are strictly zero, does not exist in nature. The frame of reference of distant fixed stars is a good approximation to an ideal inertial frame.

S43 *You have emphasized throughout the discussion that the inertial or non-inertial character of a frame can be ascertained only empirically. But surely, there must be some deeper reason why, in nature, some frames turn out to be inertial and others non-inertial. For example, in Newton's pail experiment, why is it that*

---

† We have ignored in this discussion the angular velocity of revolution of the earth around the sun.

the laboratory frame is inertial and the frame of the rotating pail is non-inertial, and not the other way?

T43 This is an important question, and I wish I could give you a definitive answer. Mach's famous principle is a response to a question of this kind. In the pail experiment, if the inertial observer looks out to the sky, he will find the distant stars fixed. On the other hand, the non-inertial observer finds the distant stars to be rotating. Thus we find that a (local) inertial frame happens to be one in which the distant stars in the universe do not rotate.

Mach converted this coincidence into a principle. He conjectured that, in a proper cosmological theory, local inertial or other character of a frame is determined by matter distribution in the entire universe. In other words, in this view the pseudo-forces in the frame of the rotating pail are actually material in origin—they arise from distant matter in the universe. I suggest you consult a more advanced text to get a proper view of these rather subtle matters.

Examples of non-inertial frames

S44 *Let us turn to more mundane matters. Could you give me some examples where it is more convenient to consider non-inertial frames?*

T44 Consider a box lying on the floor of a train that is moving with an acceleration $\boldsymbol{a}$. Can you explain the motion of the box from the viewpoint of a ground observer?

S45 *That is simple. The weight of the box is cancelled by the normal reaction on it by the floor. The horizontal acceleration of the box is caused by the force provided by the engine driving the train.*

T45 The last statement of yours is very vague physics-wise. We emphasized in Chapter 4 that in mechanics we should invoke only two kinds of material forces: the gravitational force and contact force. So, instead of making a vague layman-like statement that the engine gives a force on the box, you should say that the horizontal acceleration of the box is caused by the frictional force between the box and the floor.

S46 *What? Frictional force in the same direction as the direction of motion? I thought it always opposed motion.*

T46 Static friction opposes <u>*impending*</u> relative motion between surfaces in contact. In the absence of friction, for a ground observer, the box will stay where it is (I law), while the floor of the train will slide by. To oppose this impending relative motion, the frictional force drags the box with the same acceleration $\boldsymbol{a}$ as that of the floor. Relative to the floor, therefore the box remains at rest.

It is much easier to see this relative to the frame of reference of the accelerated train. This frame is non-inertial and therefore the box is subject to a pseudo-force $-m\boldsymbol{a}$. To oppose the impending motion due to this force, an equal and opposite static frictional force comes into play keeping the box stationary.

S47 *Any other example of a non-inertial frame in practice?*

T47 Consider the uniform circular motion of a stone tied to the end of a string, relative to the laboratory frame of reference which is given as inertial. For circular motion, kinematically, $a = v^2/r$ and it is directed inward. This centripetal acceleration is caused by a material force namely tension $\boldsymbol{T}$ in the string. Therefore, according to the II law

$$T = \frac{mv^2}{r}.$$

Notice, that nowhere have I used the word centrifugal force. People sometimes describe this situation as : 'The centrifugal force on the stone balances the centripetal force of tension keeping the stone in equilibrium'. This explanation of motion of the stone relative to the laboratory frame is wrong on two counts. First, in the laboratory frame, the stone is <u>not</u> in equilibrium. Second, a centrifugal force is a pseudo-force which should never be invoked in an inertial frame.

S48 *But the explanation is met so often; there must be something to it.*

T48 It is a correct explanation in a non-inertial frame which is rotating with the stone and has its origin at the centre of the circle. The angular velocity of this frame is constant-the same as that of the stone. Also, the Coriolis force on the stone is zero since it is at rest in this frame. Thus, the only pseudo-force on the stone is the centrifugal force and hence the equation of motion for the stone in this frame is

$$m\boldsymbol{a} = \boldsymbol{F} - m\boldsymbol{\omega} \times (\boldsymbol{\omega} \times \boldsymbol{r}).$$

Now, relative to the rotating frame the stone is in equilibrium, so that $\boldsymbol{a} = 0$. The material force continues to be the tension $\boldsymbol{T}$, whereas the magnitude of the centrifugal force is given by $m\omega^2 r$. Thus the above equation reduces to that in T47.

'Feeling' or 'experiencing' a force

S49 *You said that the centrifugal force is not a real material force. But one does experience it in practice, on a merry-go-round for example. How do you explain it?*

T49 First of all, banish any idea that pseudo-forces are not 'real' forces. In a rotating frame, for example, the centrifugal force is as real as any material force. The term pseudo-force is meant merely to emphasise the fact that it does not arise due to any material agency, but arises due to the rotation/acceleration of the frame itself.

On a merry-go-round, the centrifugal force on a man is balanced by the contact forces of friction and the reaction of the sides of the seat. These latter forces act only at the surfaces of the body in contact with the seat. Other parts of the body would fly off in the instantaneous tangential direction, like any other object let loose, if the forces constraining them were 'switched off'. The constraining forces acting on various parts of the body are generated in the muscles, bones etc., and the forces exerted by the muscles and bones etc. have their origin in the stresses developed in them. These body stresses are associated with our 'experience' of the centrifugal force. The net result is that each part of our body is subject to an inward material force which balances the outward centrifugal force.

The interesting thing is that the brain interprets the body stresses as our experience of an outward force. Thus we 'feel' the force in the direction of the impending motion.

S50 *How does a bystander view the situation?*

T50 In the ground's (inertial) frame of reference, there is no centrifugal force. The inward material force on each part of the body causes the observed circular motion in accordance with the II law.

S51 *Suppose a person is standing on a rotating disk; imagine that the friction is suddenly switched off. There is then no inward material force and he will be pushed outward by the centrifugal force. How will a bystander, watching this motion, explain it without invoking the centrifugal force?*

T51 Be careful. What is outward motion† in the rotating frame is a straight-line motion, along the instantaneous tangential direction, to a bystander on the ground. He easily explains this motion as due to inertia (the I law) and does not need to invoke any force.

S52 *Suppose I invert the problem. Relative to the person in the merry-go-round, the outsider will seem to move in a circle in the opposite sense. Now, there is no force on him because he is outside the merry-go-round. How can the person in the merry-go-round explain the circular motion of the outsider relative to him?*

T52 You have made a serious error here. I warned you early in this chapter not to think that a frame of reference is bounded and localized. The frame of reference rotating with the merry-go-round does not terminate at its boundary. It is wrong to think that in this rotating frame objects on the merry-go-round only experience pseudo-forces. In this frame all objects in the universe experience pseudo-forces.

S53 *I realize the error. But I am still confused about one thing. In the rotating frame, acceleration of the outsider is centripetal, i.e., inward, while the centrifugal force on him is outward. There seems to be a mismatch of directions.*

T53 The error lies in ignoring another pseudo-force, the Coriolis force. Go back to T28 [Eq. (20)]. If an object is stationary in a rotating frame, the centrifugal force is the only pseudo-force acting on it. But for an object that is in motion relative to a rotating frame, there is an additional velocity-dependent term: the Coriolis force. Since the outsider is in motion relative to the merry-go-round, he will be subject to both the centrifugal and Coriolis forces. It turns out, in this case, that the latter is twice in magnitude and directed inward. The net force on the outsider, therefore, is inward as you required (See Box 5.3).

† It should be noted that the outward motion is governed not only by the centrifugal force but also by the velocity-dependent Coriolis force. Thus, in the rotating frame, motion with $\boldsymbol{F} = 0$ is complicated and not purely radial as is sometimes erroneously claimed.

Box 5.3

Consider a merry-go-round rotating with a constant angular velocity $\boldsymbol{\omega}$. A person of mass m is standing on the ground. Let $\boldsymbol{r}$ denote the position vector of the person. Since the person is stationary on the ground, his velocity $\boldsymbol{v}$ relative to the ground's frame (S) is zero. Equation (9) gives his velocity $\boldsymbol{v}'$ relative to the S' frame rotating with the merry-go-round : $\boldsymbol{v}' = -\boldsymbol{\omega} \times \boldsymbol{r}$ (See Fig. 5.9).

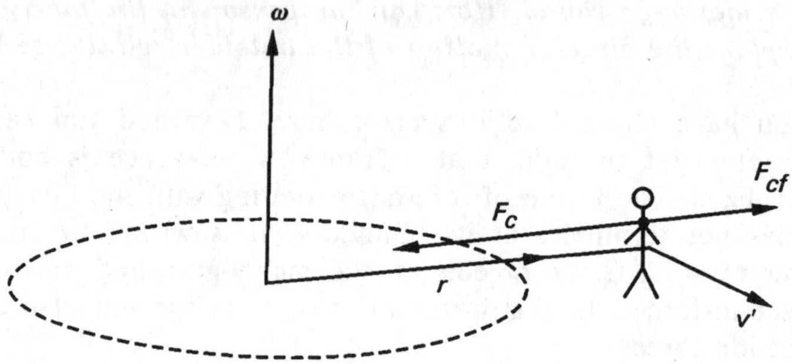

Fig. 5.9 The platform rotates with an angular velocity $\boldsymbol{\omega}$. $\boldsymbol{v}'$ is the velocity of the bystander relative to the rotating frame of the platform.

The centrifugal force on the person is $\boldsymbol{F}_{cf} = -m\boldsymbol{\omega} \times (\boldsymbol{\omega} \times \boldsymbol{r})$.

The Coriolis force on the person is $\boldsymbol{F}_c = -2m\boldsymbol{\omega} \times \boldsymbol{v}' = 2m\boldsymbol{\omega} \times (\boldsymbol{\omega} \times \boldsymbol{r})$

Clearly, the centrifugal force points away from the centre, whereas the Coriolis force points towards the centre. Also, the magnitude of the Coriolis force is twice that of the centrifugal force. Thus, the net force on the bystander has a magnitude $m\omega^2 r$, the same as the magnitude of the centrifugal force, but points inward.

S54 *You said that, relative to the rotating frame, pseudo-forces act on the man on the ground as well. Yet we know that the man does not experience any such force. How do you explain that?*

T54 I explained in T49 that the stresses developed in our body, in order to oppose any impending relative motion between different parts of our body, are associated with our 'experience' of a force. And the direction of the force 'felt' is the direction of the impending motion. Now, if the impending motion is not opposed, no stresses are produced in the body and no force is felt ! An example will clarify the point.

An astronaut in outer space far away from all bodies feels weightless because there is no force acting on him; there is no impending motion between different parts of the body and hence, no efforts on his part to oppose it.

While standing on the surface of the earth, the normal force by the ground acts on the feet of a person. Other parts of the body would fall downwards under gravity if there were no forces constraining them together. These forces have their origin in the stresses developed in the body. These stresses are associated with our 'experience' of weight. Note that the net force on each part of the body is zero; but the brain interprets the body stresses as a force in the downward direction — the direction of impending motion.

Now consider an astronaut in a satellite orbiting the earth. The gravitational force on him is not zero; in fact that is the force which keeps him in the orbit, along with the satellite and everything else in it. Why, then, does he 'feel' weightless? Because all the parts of his body are freely falling under the force with the same acceleration; there is no relative motion between the different parts that he needs to oppose. Consequently, no stresses are produced in his body and he 'experiences' no force.

Returning to your question, the fact that a bystander does not 'feel' any force (apart from his weight) can be explained both by a ground observer and a rotating observer. For a ground observer, of course, there are no pseudo-forces on the man. Relative to the rotating frame, there <u>are</u> pseudo-forces acting on the bystander. But, all parts of his body are rotating freely under their action with the same acceleration and there is no relative motion between the different parts that he needs to oppose. Hence, he does not 'feel' the pseudo-forces.

S55 *One more query. Pseudo-forces or frame-dependent forces are also called 'inertial forces'. What is the origin of this term?*

T55 There is a simple reason behind that term. As you know, the constant m appearing in Newton's II law $\boldsymbol{F} = m\boldsymbol{a}$ is called the inertial mass of the body since that is a measure of the inertia of the body (the property of resisting change in its state of motion). This <u>same</u> constant m also appears in pseudo-forces in a non-inertial frame. The constants appearing in the material forces may, however, be different. In particular, the mass constant that appears in the gravitational force is the gravitational mass of the body. It is another matter that the gravitational mass turns out to be proportional (or equal by choice) to the inertial mass of a body. If the body is charged, material forces may include other constants like charge, magnetic moment, etc. But whatever the constants appearing in material forces, the pseudo-forces always contain only the inertial mass constant of the body. It is to highlight this fundamental difference between material forces and pseudo-forces that the latter are called 'inertial forces'.

Beware of possible terminological confusion here. The term 'pseudo-forces', as we saw, can give you a wrong impression that they are fictitious, not real, even in a non-inertial frame. The term 'inertial forces' can sometimes erroneously suggest that they are forces in an inertial frame! The term 'frame-dependent forces' is perhaps the least confusing of all.

Relativity of Einstein

S56 *I have a final question on these matters. We have talked of relativity in Newtonian mechanics. In what ways does it differ from the relativity of Einstein?*

T56 I can only answer this qualitatively. A detailed discussion of Einstein's relativity is beyond the scope of this book. As we have seen, Newtonian theory satisfies the following principle of relativity:

The laws of <u>mechanics</u> are the same for different inertial observers in uniform relative motion.†

Note the emphasized word: mechanics. The basic theory of electromagnetism had not yet arrived. When it arrived two centuries later, a perplexing thing happened. Maxwell's equations of electromagnetism contained a constant that was identified with the velocity of light in a vacuum. Now in Newtonian mechanics, a velocity cannot have the same value in two frames in relative motion. Thus the laws of electromagnetism, involving one definite value of velocity, seemed to pertain to one particular frame. That is, the laws did not satisfy Newtonian relativity, and seemed to single out a privileged frame among the class of inertial frames, which could be taken to be the frame of absolute rest: ether.

In short, it seemed that the laws of mechanics did not require ether, while the laws of electromagnetism did. Einstein was not willing to accept this unsymmetric state of affairs. †† He rejected the idea of ether and postulated that the laws of physics, whether of mechanics or of electromagnetism are identical for all inertial

† This implies that the laws do <u>not</u> require a frame of absolute rest. Newton was aware of this; yet, paradoxically, he still believed in a frame of absolute rest — the ether. The ether was thought of as something that mediated the gravitational force. Other reasons may be partly aesthetic and partly a product of the time: it was difficult in those times to imagine a cosmic order based on equity !

†† Also, careful experiments to detect ether had yielded negative results. However, it is not very clear whether Einstein was aware of these, or to what extent they motivated his ideas.

observers. This is Einstein's first postulate: the principle of (special) relativity:

(I) The laws of physics are the same for all inertial observers in uniform relative motion.

Now if you believe Maxwell's equations to be the correct laws of electrodynamics, postulate (1) implies that the constant c in these equations, i.e. the velocity of light should be the same relative to all inertial frames. This conclusion runs counter to our common sense. It violates the usual velocity addition rule that we use so often:

$$\boldsymbol{v} = \boldsymbol{v}' + \boldsymbol{V}.$$

S57 *Then how did Einstein get around the problem? How did he explain the constancy of the velocity of light relative to different frames?*

T57 He did not. His genius lay in turning the problem around. He took the constancy of c as another postulate of his theory:

(II) The velocity of light in vacuum is the same for all inertial frames, i.e. it is independent of the relative motion between the source and the observer,

and he used it to modify the velocity addition rule so that the latter is consistent with (II)! Now the velocity addition rule is based on our very elementary and intuitive ideas of of distance, time, etc.—ideas so basic that nobody ever thought of questioning them. So Einstein was led to alter our conceptions of space and time.

S58 *I have heard that last statement so often. What is it supposed to mean?*

T58 Let us first spell out our intuitive conceptions of space and time. First, we feel that the time of an event has little to do with the external observer, Of course, different observers may differ in the zero settings of their clocks. So, strictly, we should say that the time difference between two events is independent of the frame of reference. In particular, if two events are simultaneous to one observer, they will be so to any other observer. Likewise, the dimensions of an object, e.g. the length of a rigid rod seems to be its intrinsic property that has nothing to do with its motion, i.e. it is independent of the frame of reference. These intuitive ideas underlie the Galilean transformations [Eq. (6)] that we obtained early in this chapter.

S59 *So what is wrong with these intuitive ideas?*

T59 We have seen that the velocity addition rule, in particular, is a corollary of Galilean transformations. Since this, as we saw, is inconsistent with Einstein's theory, he had no choice but to reject Galilean transformations, and replace them by new transformations, based on the constancy of velocity of light relative to all inertial frames. These new transformations are known as Lorentz transformations. I quote them without proof in Box 5.4. You will need to consult a standard text on relativity to understand them properly.

Box 5.4

Lorentz Transformations:

Let S be an inertial frame and let S' be another frame moving uniformly relative to S with a velocity v along say the x-axis of S. Let the axes of the S' frame be parallel to those of S as shown in Fig. 5.10. Let O and O' coincide at $t = t' = 0$.

Suppose that an event occurs at (x, y, z) at time t relative to the S frame. Relative to the S' frame, the same event occurs at time t' at (x', y', z'). The space–time coordinates of the same event relative to the two frames are related to one another by the equations:

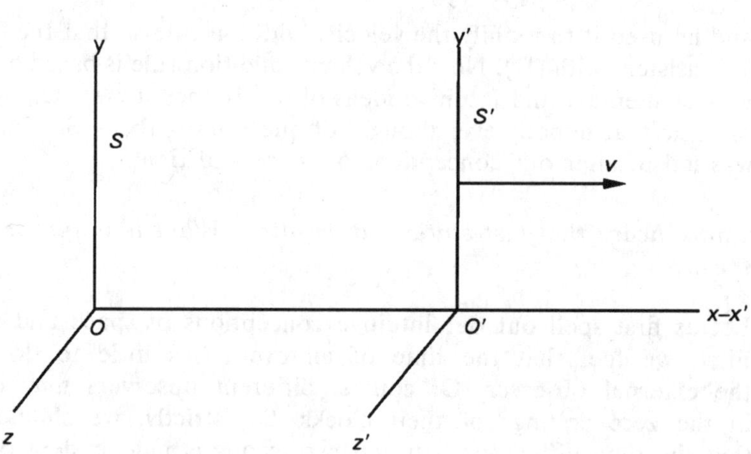

Fig. 5.10 The frame S' moves with a uniform velocity v relative to S along the common $x - x'$–axis.

$$x' = \frac{x - vt}{\sqrt{1 - v^2/c^2}}$$

$$y' = y; \quad z' = z, \qquad t' = \frac{t - vx/c^2}{\sqrt{1 - v^2/c^2}}.$$

These relations are known as Lorentz transformations. Four important kinematic implications of the Lorentz transformations are worth keeping in mind:

1) Relativity of simultaneity:

If two events (at different locations) are simultaneous in a frame S, they are not so in a frame S' in relative motion. Simultaneity, therefore, is no longer absolute as in Newtonian relativity, but is a relative concept.

2) Length contraction:

Length determination, as we know, requires simultaneous location measurements. But if two observers do not agree on simultaneity, they will naturally not agree on their measured lengths. If ℓ_0 is the length of a rigid rod measured when it is at rest relative to an observer, and ℓ its length while it is moving along the direction of its length with a velocity v (relative to the same observer), then $\ell = \ell_0\sqrt{1 - v^2/c^2}$. In other words, the length of a moving rigid rod is shorter than its length at rest.

3) Time dilation:

This refers to the time interval Δt between two events at the same location in one frame (S). In another frame (S') which is moving uniformly relative to the first, the time interval between the same events, $\Delta t'$, is measured and found to be greater:

$$\Delta t' = \frac{\Delta t}{\sqrt{1 - v^2/c^2}}.$$

4) Velocity addition rule:

The ordinary velocity addition rule is no longer true; the relativistic velocity addition rule ensures that c is the same for all frames, as it should be for consistency, since that is the starting axiom employed in the derivation of the Lorentz transformations.

If u is the velocity of an object relative to S, and u' the velocity of the same object relative to S', then

$$u' = \frac{u - v}{1 - uv/c^2}.$$

You can check that for $u = c$, $u' = c$.

S60 *Do Newtonian laws of motion remain unchanged in special relativity?*

T60 You must get this clear.

Newtonian laws of motion are invariant, i.e. they retain the same form under Galilean transformations, but not under the Lorentz transformations.

The laws of electromagnetism (Maxwell's equations) are invariant under the Lorentz transformations, but not under the Galilean transformations.

So, if the laws of physics are to be consistent with special relativity based on the Lorentz transformations, Maxwell's equations do not require any modification but Newton's laws of motion do.

S61 *What are these modified laws of motion in special relativity?*

T61 Newton's first law of motion is the same as before; it just defines an inertial frame. The II law of motion in the form

$$\boldsymbol{F} = \frac{d\boldsymbol{p}}{dt} \tag{21}$$

is still true, but the relativistic definition of momentum must be carefully noted:

$$\boldsymbol{p} = \frac{m_0\boldsymbol{v}}{\sqrt{1 - v^2/c^2}}. \tag{22}$$

One way of interpreting this new definition is to lump m_0 and $1/\sqrt{1 - v^2/c^2}$ together and call the product as the 'moving mass' of the particle

$$m = \frac{m_0}{\sqrt{1 - v^2/c^2}}. \tag{23}$$

If v in this equation is zero, $m = m_0$. m_0 is, therefore, called the rest mass of the particle. Equation (23) is the basis of the often-quoted phrase: mass of a body increases with speed. The important thing to note is that Eqs. (21) and (22) give

$$\boldsymbol{F} = \frac{d}{dt}\left(\frac{m_0\boldsymbol{v}}{\sqrt{1 - v^2/c^2}}\right),$$

i.e. the II law in the form $\boldsymbol{F} = m\boldsymbol{a}$ is <u>not</u> correct in special relativity.

In particular, note that the force is not necessarily in the same direction as $d\boldsymbol{v}/dt$.

To sum up, the II law in the form $\boldsymbol{F} = d\boldsymbol{p}/dt$ is true in both Newtonian and relativistic mechanics. In the former $\boldsymbol{p} = m_0\boldsymbol{v}$, while in the latter $\boldsymbol{p} = m\boldsymbol{v}$ where m is given by Eq. (23).

S62 *What about the III law in relativity?*

T62 The situation is a bit more involved in regard to the III law. In the Newtonian picture, the mutual forces between bodies (e.g., gravitational force, coulomb force etc.) located any distance apart are exerted instantaneously. This is against special relativity (and fact), since no signal can travel faster than the speed of light. Thus in a proper relativistic domain, e.g. for the mutual force between electrons moving at high speeds, the III law is not valid. Of course, for ordinary distances and speeds the times taken by signals to communicate forces are negligible and the III law is approximately valid.

For contact forces, obviously, there is no problem of the above sort, and we can continue to use the III law in relativistic mechanics.

6

Energy and Momentum

Work–Energy theorem

S1 *In physics books we often see the statement: "energy is the capacity to do work." What is this supposed to mean?*

T1 To understand the phrase, it is not much use pondering over its literal meaning; let us look at the physical idea behind it.

You know that the final velocity of a body, under the action of a force depends on the initial velocity of the body, the magnitude and direction of the force and the time interval or the distance over which the force, acts on the body. If the quantity of interest is the magnitude of the final velocity, we can then do the following:

If the force is constant, we combine the II law $\boldsymbol{F} = m\boldsymbol{a}$ with the kinematical relation $v_f^2 - v_i^2 = 2\boldsymbol{a}.\boldsymbol{s}$, valid for constant acceleration, to obtain

$$\boldsymbol{F} \cdot \boldsymbol{s} = m\boldsymbol{a} \cdot \boldsymbol{s} = \frac{1}{2}m(v_f^2 - v_i^2).$$

More generally, if $\boldsymbol{F}$ is varying, we have

$$\begin{aligned} \boldsymbol{F} \cdot d\boldsymbol{s} &= m\frac{d\boldsymbol{v}}{dt} \cdot \boldsymbol{v}dt \\ &= \frac{m}{2}d(\boldsymbol{v} \cdot \boldsymbol{v}) \\ &= d\left(\frac{1}{2}mv^2\right), \end{aligned}$$

which may be integrated as follows:

$$\begin{aligned} \int_i^f \boldsymbol{F} \cdot d\boldsymbol{s} &= \int_i^f d\left(\frac{1}{2}mv^2\right) \\ &= \frac{1}{2}m\left(v_f^2 - v_i^2\right). \end{aligned}$$

The notions of work and kinetic energy are introduced at this point. The right hand side of the equation is a change in the quantity: half the mass times the square of the speed of the body. We call this the kinetic energy of the body and denote it by T. The left-hand side is called the work done by the force $\boldsymbol{F}$ on the body over its path from i to f, and is denoted by W. We then write the above equation as

$$W = T_f - T_i.$$

This is known as the work–energy theorem. It says that the work done by the net external force on a body equals the change in its kinetic energy. As we have seen, this theorem is a direct consequence of the II law of motion.

S2 *Does that mean I can always use the work–energy theorem in place of the II law, while solving problems in mechanics?*

T2 The theorem is based on the II law; but notice that the II law is a vector equation (which is equivalent to having three scalar equations), while the work–energy theorem is a single scalar equation. Thus, the latter does not have the full dynamical information contained in the II law. So, when you can solve the equation $\boldsymbol{F} = m\boldsymbol{a}$ for a problem, you have all the possible information about the motion. Sometimes, however, solving the equation of motion may be difficult or unnecessary. Important information about the motion, in such situations, can be obtained from energy considerations alone (See S30–T30).

S3 *I am clear about the fact that work gets converted into kinetic energy. But in what sense is kinetic energy the 'capacity' to do work?*

T3 Suppose a block of mass m is moving with velocity v, so its kinetic energy is $\frac{1}{2}mv^2$. Suppose, further, that it hits an elastic spring clamped on a wall. As the spring gets compressed, it exerts a force F_{bs}, on the block. The block exerts an equal and opposite $F_{sb} = -F_{bs}$ on the spring. The velocity of the block decreases progressively because of the retarding force $\boldsymbol{F}_{bs}$, finally (momentarily) becoming zero. The work W_{sb} done by the block, during the course of retardation, on the spring is given by

$$\begin{aligned} W_{sb} &= \int_i^f \boldsymbol{F}_{sb} dx \\ &= -\int_i^f \boldsymbol{F}_{bs} dx = -(T_f - T_i), \end{aligned}$$

where T_i and T_f refer to the initial and final kinetic energy of the block, respectively. Since $T_i = \frac{1}{2}mv^2$ and $T_f = 0$, the work done by the block before coming to rest is $W_{sb} = \frac{1}{2}mv^2$

Thus, the kinetic energy is indeed the amount of work that an object can do before coming to rest.

When is the work of a force zero?

S4 *It is always emphasized that a force works only when the point of application of the force moves. Yet when I press a wall with some force, or hold an object in my hand, I do seem to do work; else why should I feel tired after some time?*

T4 You must clearly distinguish between the technical meaning of the term 'work' in physics and the meaning of this word in ordinary language. Physics-wise, the force on the wall does no work, since the wall does not move. Likewise, the force due to gravity, or the reaction on the object by the hand, does not do any work if the object is stationary (In practice, it keeps shaking slightly).

What then explains the feeling of tiredness? You see, our body is not rigid; it is a complex system which is deformable within certain limits. When you press against a wall, or provide reaction to an object in your hand, the body configuration is altered slightly from what it is in the relaxed state. To bring and sustain it in this altered state, certain complex processes (chemical reactions) take place in your body, resulting in the depletion of nutrients and production of heat. This causes tiredness.

Anthropomorphic† considerations are a very common source of confusion in understanding the basic concepts of mechanics. To avoid this pitfall, in particular problems, it is best to replace humans by mechanical devices! The device can be arranged to provide the same force on the external object as the human being does. The physics regarding the external object remains the same. And the internal details of the force agency, which are irrelevant for the problem, do not mislead you.

Thus, the force on the wall provided by you can be modelled by say a suitable arrangement of rigid rods and springs. It is then seen that the system of rods and springs provides a force but, as long as the configuration is static, does no work. In providing the force, stresses are developed in the mechanical assembly. However, the state of the assembly, in the stressed condition is an equilibrium

† human-like, pertaining to humans.

state and, unlike in the case of human body, no energy need be expended in order to keep it in that state.

S5 *Is it possible that there is a net force on a body in motion, and yet no work is done by the force?*

T5 Yes, of course. Whenever a force is normal to the displacement of the body, the force does no work. In uniform circular motion there is a net force on the body in the radially inward direction, whereas the displacement at every point is in the tangential direction. Hence, no work is done by the force and the kinetic energy of the body is constant.

S6 *Is circular motion the only example of this type?*

T6 No. All that is required is that at every instant the force $\boldsymbol{F}$ be normal to the velocity at that instant. An important example is the motion of a charged particle in a magnetic field. The magnetic force on a charge is given by

$$\boldsymbol{F} = q\boldsymbol{v} \times \boldsymbol{B}.$$

Clearly, the magnetic force is always perpendicular to the velocity and, hence, it does no work.

The fact that the force is always perpendicular to velocity, by no means implies that the trajectory of the particle is necessarily circular. It is circular if the magnetic field is uniform and the initial velocity of the charged particle is normal to the field. In case the initial velocity has a component parallel to the direction of the uniform magnetic field, that component remains unaffected by the field. The net effect is that the particle has a uniform translational motion in the direction of the field, superposed on a uniform circular motion in a plane normal to the field. The motion of the particle, then, is not circular but helical. For a non-uniform magnetic field the trajectory can be even more complicated. The work done by the magnetic force, nevertheless, is zero in every case.

In the above examples, the work over <u>any</u> segment of the path of the particle is zero. We shall see later that for the so-called conservative forces the work over an arbitrary path is not zero, but that over a closed path is zero.

S7 *Is it possible that the net external force on a body is zero, and yet some work is done on it?*

T7 For a single point particle this is obviously not possible, but for extended bodies, or in general for a system of particles, this is possible.

For a rigid body, for example, the net external force may be zero and yet the net torque may not be zero. The torque does work over the angular displacement of the body, thereby changing the rotational kinetic energy of the body.

For a deformable body, the points of application of different forces on the body may move through different displacements. Thus it is possible that even if $\sum \boldsymbol{F}_i = 0$, we have $\sum \boldsymbol{F}_i \cdot \Delta \boldsymbol{s}_i \neq 0$. A familiar example is a spring compressed by two equal and opposite forces at its ends. Clearly, the net force on the spring is zero, but the net work done by the forces is not zero.

Note that if the net force is zero, the acceleration of the centre of mass of the system is zero and thus there is no change in the translational kinetic energy of the centre of mass of the system.

S8 *I have another query. Is the work done by A on B equal in magnitude but opposite in sign to the work done by B on A?*

T8 The III law of Newton says that the force on A by B is equal and opposite to the force on B by A. There is no similar statement possible for the quantity: work. Take care not to be misled by this pseudo-III law for work. Let us get this clear by an example.

Consider the fall of a stone under the gravity of the earth. The earth exerts a force mg on the stone; the stone exerts an equal and opposite force on the earth (III law). However, the earth is much more massive than the stone. So, for the same magnitude of force, its displacement is negligible compared to that of the stone. Thus the work done by the earth on the stone is mgh, where h is the displacement of the stone; but the work done by the stone on the earth is negligible.†

S9 *I understand your point. But one does seem to use this kind of rule at many places. For example, in thermodynamics one often says that the work done by the system on the surroundings is the negative of the work done on the system by the surroundings.*

T9 That statement is correct in the context in which it is made. Consider a gas in a cylinder with a movable piston. Suppose the piston is

† The (negligible) displacement of the earth, in fact, is in the same direction as the force on it due to the stone. The work done by the stone on the earth is, thus, positive. The work done by the earth on the stone is, likewise, positive. The quantities, clearly, are neither equal in magnitude nor opposite in sign.

pushed down with some acceleration. In this case, the work done by the outside agency on the piston is positive, while the work done by the system (gas) on the piston is negative. The two quantities are not equal in magnitude; their sum, in fact, equals the increase in the kinetic energy of the piston, according to the work–energy theorem.

In equilibrium thermodynamics, however, the piston is imagined to move quasi-statically. That is, at every stage the piston's kinetic energy is zero. In this case the two works referred to above are negative of each other.

Universality of work–energy theorem

S10 *Is the work–energy theorem true in non-inertial frames?*

T10 Yes, provided we take into account the work done by the pseudo-forces also. Take, for simplicity, a non-inertial frame that has only translational acceleration $\boldsymbol{\alpha}$. In this frame the II law takes the form

$$m\frac{d\boldsymbol{v}}{dt} = \boldsymbol{F} - m\boldsymbol{\alpha}.$$

Following the steps in T1, we obtain the work–energy theorem in a non-inertial frame as

$$\frac{1}{2}m(v_f^2 - v_i^2) = \int_i^f \boldsymbol{F}\cdot d\boldsymbol{r} - m\int_t^f \boldsymbol{\alpha}\cdot d\boldsymbol{r}.$$

Thus, in a non-inertial frame, the kinetic energy of an object changes not only due to the work done by the material forces but also by the work of the pseudo-forces. In particular, even in the absence of material forces, the kinetic energy of an object changes due to the latter.

S11 *Is the work–energy theorem true in special relativity?*

T11 Yes, in a suitably generalized form. As we saw in Chapter 5, the II law in the relativistic domain is

$$\boldsymbol{F} = \frac{d}{dt}(m\boldsymbol{v}). \tag{1}$$

where

$$m = \frac{m_0}{\sqrt{1 - v^2/c^2}}. \tag{2}$$

Multiplying by $\boldsymbol{dr} = \boldsymbol{v}dt$ on the two sides of Eq. (1), we write

$$\boldsymbol{F} \cdot \boldsymbol{dr} = \frac{d}{dt}(m\boldsymbol{v}) \cdot \boldsymbol{v}dt = \left(\frac{dm}{dt}\boldsymbol{v} \cdot \boldsymbol{v} + m\frac{d\boldsymbol{v}}{dt} \cdot \boldsymbol{v}\right) dt \tag{3}$$

Now, Eq. (2) may be written in the form

$$m^2c^2 = m^2v^2 + m_0^2c^2.$$

Differentiation of this equation with respect to time gives

$$\frac{dm}{dt}c^2 = \frac{dm}{dt}v^2 + m\boldsymbol{v} \cdot \frac{d\boldsymbol{v}}{dt}$$

Substituting this in Eq. (3), and integrating from the initial state i to the final state f, we get

$$\int_i^f \boldsymbol{F} \cdot \boldsymbol{dr} = m_f c^2 - m_i c^2. \tag{4}$$

Since the left hand side is the work done, it may be natural to identify the right hand side as the change in kinetic energy $T_f - T_i$. It will be erroneous, however, to take kinetic energy $T = mc^2$. We expect the kinetic energy of a body to be zero when its velocity is zero whereas $mc^2 \to m_0c^2$ as $v \to 0$. Thus, in special relativity, kinetic energy is defined as

$$T = mc^2 - m_0c^2.$$

This suggests that we introduce a new physical quantity called total energy E defined as

$$E \equiv mc^2 = T + m_0c^2.$$

The total energy E includes the kinetic energy T and the so-called rest energy m_0c^2.

The work–energy theorem in relativity [Eq. (4)] is expressed in terms of E as

$$W = E_f - E_i.$$

S12 *What is the significance of this? Is $E_f - E_i$ not the same thing as $T_f - T_i$?*

T12 Ordinarily the rest mass m_0 is a constant and $E_f - E_i$ is the same as $T_f - T_i$. However there are situations where this does not hold true. And that is when we come to a most dramatic consequence of special relativity.

Suppose two bodies, each of rest mass m_0, collide head-on with each other with equal and opposite velocities relative to a particular frame of reference S. Clearly, relative to S, the total momentum before collision is zero:

$$\boldsymbol{P}_{\text{before}} = 0.$$

Also, the kinetic energies, and hence the total energies, of the two bodies are equal. If $T + m_0c^2$ is the energy of one body, the total energy of the system before collision is

$$E_{\text{before}} = 2(T + m_0c^2).$$

Suppose, now, that the collision is perfectly inelastic so that the two bodies combine to form a single composite. Momentum conservation requires that the net momentum after collision be zero. Hence the velocity, and the kinetic energy, of the composite body must be zero. Total energy conservation then requires that

$$E_{\text{after}} = 2(T + m_0c^2).$$

Clearly, the mass M_0 of the composite body must be such that

$$M_0c^2 = 2(T + m_0c^2),$$

Evidently,

$$M_0 > 2m_0.$$

We thus see that kinetic energy of the colliding bodies appears as increased rest mass energy of the composite body. This is the well-known interconvertibility of mass and kinetic energy in relativistic dynamics.

Path independence of work and potential energy

S13 *Let us turn to the notion of potential energy. It is supposed to be the energy associated with the position of a system or its configuration. What does that mean?*

T13 Before I come to it, let me emphasize one thing. The definition of kinetic energy makes no reference to forces acting on an object. Likewise, the notion of work done by a force is general; it makes no reference to the nature of the force. Consequently, the work-energy theorem is true, no matter what kind of forces act on the body.

The notion of potential energy, on the other hand, is applicable only to a class of forces called *conservative forces*. (The reason for naming them 'conservative' will be clear to you when we go up to T18) Thus, for example, one can associate potential energy with gravitational force, Coulomb force, spring force etc., but not with friction or viscous forces etc.

S14 *I am surprised. What is this special class of forces?*

T14 Let me explain it with an example. Take the gravitational force that you are familiar with. This force has one very fundamental property that you must know. It is the following:

Consider the displacement of a particle from point 1 to point 2 in space under the influence of the gravitational force. It can be shown that the work done by the gravitational force on the particle depends only on the initial and final positions of the particle; it is independent of the path taken by the particle between the points 1

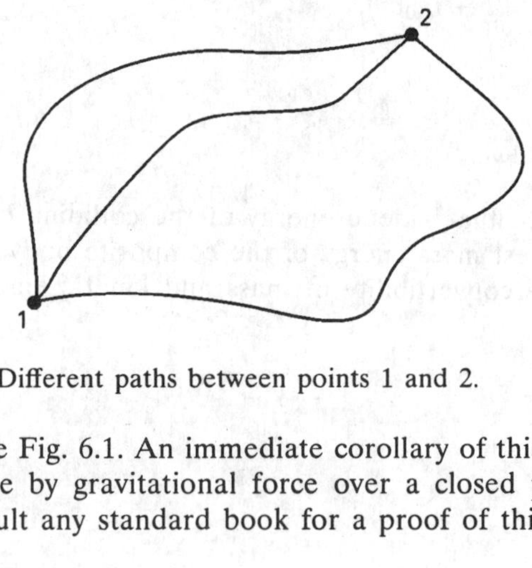

Fig. 6.1 Different paths between points 1 and 2.

and 2. See Fig. 6.1. An immediate corollary of this result is that the work done by gravitational force over a closed path is zero. You may consult any standard book for a proof of this result.

Any force for which a similar result holds true, i.e. any force for which work is independent of the path, is called a conservative force. Gravitational force, Coulomb force and elastic spring force are examples of conservative forces. Friction is a non-conservative force. The notion of potential energy can be introduced for conservative forces only.

S15 *I am a little confused about this 'path-independence'. A particle under the action of a given force follows a definite path depending upon the initial conditions. What does one mean by taking different paths of the particle between two points? An arbitrary path may not be a possible path of the particle for a given force. Isn't that so?*

T15 What you must realize is that, although we are interested in the work done on a particle by a particular force, that may not be the only force acting on the particle. For example, a stone can be dropped from the top of a building to a point on the ground, or a person could carry it down via a winding staircase. The first path of the particle is under the gravitational force only; the second path is possible because the particle is subject to both the gravitational force and the appropriate force provided by the person. The point is that the work done by the gravitational force is the same for the two paths, or for any other path that you take between the same two locations.

More formally, if you prefer, the path independence may simply be taken as a mathematical property of the integral $\int_1^2 \boldsymbol{F} \cdot d\boldsymbol{r}$ that follows from the particular form of the force.

S16 *I grant you this 'path-independence of work' property of gravitational force. What is so signiticant about it?*

T16 It is precisely this property that leads to the notion of potential energy. It can be proved that the work done on a body of mass m by the gravitational force due to a mass M at the origin is given by

$$\int_1^2 \boldsymbol{F} \cdot d\boldsymbol{r} = -G\frac{Mm}{r_1} + G\frac{Mm}{r_2}.$$

Note that the right hand side depends only on the initial and final locations of mass m and not on the intervening path. If we define a quantity V that depends only on position by

$$V(\boldsymbol{r}) = -G\frac{Mm}{r},$$

then

$$\int_1^2 \boldsymbol{F} \cdot d\boldsymbol{r} = V(\boldsymbol{r}_1) - V(\boldsymbol{r}_2).$$

Thus the work done by the gravitational force on a particle during its displacement from $\boldsymbol{r}_1$ to $\boldsymbol{r}_2$ equals the difference between the values of the quantity V at the points $\boldsymbol{r}_1$, and $\boldsymbol{r}_2$. We can give this quantity some name: it is called 'gravitational potential energy'.

S17 *Why is it called 'potential energy'? In what sense is it energy? The way you have arrived at it, it is just some position-dependent quantity.*

T17 I have stressed often that in physics one should not worry much over what a particular thing is called; rather one should concentrate on the precise definitions of quantities and any facts of physical significance. Many times the names of quantities could be misleading. In this particular case, however, the name 'potential energy' makes good sense.

First note the obvious. $V(\boldsymbol{r})$ has the correct dimensions, the same as those of work, to qualify as energy. Next, if the work done by the gravitational force, in displacing the particle from $\boldsymbol{r}_1$ to $\boldsymbol{r}_2$ is positive, V decreases $[V(\boldsymbol{r}_2) < V(\boldsymbol{r}_1)]$. If, on the other hand, an external agency displaces the particle against the gravitational force, so that the work done by the gravitational force is negative, the quantity V increases $[V(\boldsymbol{r}_2) > V(\boldsymbol{r}_1)]$. In other words, work done by the external agency against the gravitational force goes to increase the quantity V. Hence V does seem to be some kind of 'stored-up energy'.

We shall see later that during motion under a gravitational force, the loss of V appears as gain in the kinetic energy T of the particle. The qualifier 'potential' is quite clear. $V(\boldsymbol{r})$ is a kind of (location-dependent) stored-up energy of the particle which has a potential of converting itself into kinetic energy (or work) when the particle is let go under the gravitational force.

S18 *I have a small query here. Why did you put a negative sign in your dejtnition of $V(\boldsymbol{r})$? What is wrong in defining $V(\boldsymbol{r}) = +(GMm)/r$?*

T18 You will appreciate it when I prove an important result:
For any kind of force, according to the work–energy theorem:

$$\int_1^2 \boldsymbol{F} \cdot d\boldsymbol{r} = T_2 - T_1.$$

Remember, in this integral, F is the <u>net</u> force on the body. For gravitational force, we have seen that

$$\int_1^2 \boldsymbol{F} \cdot d\boldsymbol{r} = V_1 - V_2.$$

If the gravitational force is the only force acting, the two equations above give

$$T_1 + V_1 = T_2 + V_2, \quad \text{i.e.} \quad T + V = \text{constant.}$$

Now, about your query regarding the sign. Of course, one could define $V(\boldsymbol{r})$ as $+(GMm)/r$. Then, obviously, we would get $T - V =$ constant. Nothing wrong. We could have called $T - V$ as net energy or something. But you will agree, it is much more cute to define V, as we have done, so that we get $T + V =$ constant. That enables us to introduce a quantity $E = T + V$, and call it the total energy of the particle. The result above shows that the total energy thus defined remains constant in the actual motion of a particle under gravitational force.

Now we see why gravitational force is called a 'conservative' force. For this force, there is a quantity (called potential energy of the particle) whose sum with kinetic energy is conserved during motion under the force.

S19 *Does the discussion above apply to any conservative force?*

T19 Yes, it does. As already said, a force is conservative if $\int_1^2 \boldsymbol{F} \cdot d\boldsymbol{r}$ is independent of the path between the points $\boldsymbol{r}_1$, and $\boldsymbol{r}_2$, i.e. its value depends only on the positions $\boldsymbol{r}_1$ and $\boldsymbol{r}_2$. This shows that, for a conservative force, there exists a scalar function $V(\boldsymbol{r})$ such that

$$\int_1^2 \boldsymbol{F} \cdot d\boldsymbol{r} = V(\boldsymbol{r}_1) - V(\boldsymbol{r}_2). \tag{5}$$

As in T18, this can be combined with the work–energy theorem to obtain

$$E = T + V = \text{constant} \tag{6}$$

for motion under any conservative force. The earlier discussion on gravitational force is a particular example of this general argument with

$$V(\boldsymbol{r}) = \frac{-GmM}{r}.$$

Other conservative forces have different forms for $V(\boldsymbol{r})$.

S20 *I don't see how you have jumped to the particular form on the right side above in Eq. (5). Why not say* $V(\boldsymbol{r}_1)+V(\boldsymbol{r}_2)$ *or* $[V(\boldsymbol{r}_1)-V(\boldsymbol{r}_2)]^3$*?*

T20 Obviously, if the path is traversed in the reverse direction, the integral should change sign. So $V(\boldsymbol{r}_1)+V(\boldsymbol{r}_2)$ will not do. Further, path-independence also implies

$$\int_1^2 \boldsymbol{F}\cdot d\boldsymbol{r} + \int_2^3 \boldsymbol{F}\cdot d\boldsymbol{r} + = \int_1^3 \boldsymbol{F}\cdot d\boldsymbol{r}.$$

Your cubed form will not satisfy this requirement. The only form possible is $V(\boldsymbol{r}_1)-V(\boldsymbol{r}_2)$. [The form $V(\boldsymbol{r}_2)-V(\boldsymbol{r}_1)$ is not really different – it merely amounts to a redefinition of V as $-V$]. So, there is no loss of generality in Eq. (5) above.

A direct proof of Eq. (5) can be given using the Stokes' theorem of vector calculus; see any standard text.

S21 *In practice, how does one check if a force is conservative or not? Must we calculate the work done by the force over an arbitrary path between two points and check its path independence?*

T21 We have defined a conservative force in terms of its integral property (namely path-independence of work). This is a global criterion as we say in physics. An equivalent local criterion (i.e. a condition at every point) is possible and is easy to work with. Consider an infinitesimal displacement $d\boldsymbol{r} = dx\hat{i} + dy\hat{j} + dz\hat{k}$. The work done by a force $\boldsymbol{F}$ over this displacement is

$$\boldsymbol{F}\cdot d\boldsymbol{r} = F_x dx + F_y dy + F_z dz.$$

If the force is conservative, Eq. (5) gives

$$\begin{aligned}\boldsymbol{F}\cdot d\boldsymbol{r} &= -dV \\ &= -\left(\frac{\partial V}{\partial x}dx + \frac{\partial V}{\partial y}dy + \frac{\partial V}{\partial z}dz\right),\end{aligned}$$

which gives

$$\boldsymbol{F} = -\left(\frac{\partial V}{\partial x}\hat{i} + \frac{\partial V}{\partial y}\hat{j} + \frac{\partial V}{\partial z}\hat{k}\right);$$

i.e. $\boldsymbol{F} = -\nabla V$. Indeed you can take this to be a definition of a conservative force: a force is conservative if it can be expressed as the gradient of a scalar function. The 'path-independence of work' follows from this definition. Note, that the choice of the negative sign in $\boldsymbol{F} = -\nabla V$ has the same reason as explained in T18. In one dimension, $F = -dV/dx$.

S22 *It is said that the absolute value of potential energy has no physical meaning. How do you see that?*

T22 That is obvious from the way we introduced the quantity $V(\boldsymbol{r})$ in Eq. (5). The left side of this equation, denoting the work, is a definite physical quantity; the force $\boldsymbol{F}$ and the points $\boldsymbol{r}_1$ and $\boldsymbol{r}_2$ determine its value completely. On the right side we have the *difference* between the value of $V(\boldsymbol{r})$ at the two points. I am, therefore, free to add any constant to the potential energy function $V(\boldsymbol{r})$; $V(\boldsymbol{r}_1) - V(\boldsymbol{r}_2)$ will still be the same. This means, $V(\boldsymbol{r})$ is undetermined to within an arbitrary additive constant. That is, the absolute value of potential energy at any one location is a matter of choice, and hence cannot have any physical significance. But once I choose some value for potential energy at any one point, the value of potential energy at all other points gets fixed; the difference in potential energy at any two points is not a matter of choice but is determined by the given force.

All this is clear from the relation $\boldsymbol{F} = -\nabla V$ also. ∇V and $\nabla (V + \text{constant})$ are the same. Clearly, given a force $\boldsymbol{F}$, V can be determined only to within a constant.

Lastly, note that there is no such arbitrariness in the definition of kinetic energy. Because of V, however, the total energy $E = T + V$ is also arbitrary to within an additive constant.

S23 *The discussion on conservative forces is now clear to me. Can you consolidate it by additional examples?*

T23 Consider first the simple one-dimensional example of the spring force $F = -kx$, where x is the extension of the spring. One way of seeing the conservative nature of the force is to obtain an appropriate potential energy function. It is easily seen that $V = \frac{1}{2}kx^2$ is the required function. (Use of this V in equation $F = -dV/dx$ leads to $F = -kx$.) We may add an arbitrary constant C to the potential energy; $V = \frac{1}{2}kx^2 + C$ also leads to the same force F. The arbitrariness in the value of potential energy may be removed by choosing a particular value for C. A convenient choice is $C = 0$, which makes $V = 0$ at $x = 0$. This choice of C, thus, corresponds to choosing zero of the potential energy at $x = 0$.

The electrostatic force between two charges, as given by Coulomb's law, is another example of a conservative force. It can be verified that the potential energy function

$$V(\boldsymbol{r}_1, \boldsymbol{r}_2) = \frac{1}{4\pi\epsilon_0}\frac{q_1 q_2}{r_{12}} + C,$$

where $r_{12} = |\boldsymbol{r}_1 - \boldsymbol{r}_2|$, leads to the vector form of Coulomb's law. † Again, the potential energy is arbitrary within an additive constant C and the arbitrariness is removed by choosing the zero of the potential energy at a convenient point. Clearly $r_{12} = 0$ is not a proper choice since $V \to \infty$ as $r_{12} \to 0$. If we choose $C = 0$, $V \to 0$ as $r_{12} \to \infty$. That is, we choose the electrostatic potential energy to be zero when the two point charges are infinitely separated.

In fact the choice of zero potential energy at infinite separation is customary and sensible for all the forces which tend to zero as the separation between the interacting bodies tends to infinity.

S24 *Is it correct to say that if the force between two bodies is attractive, the sign of the potential energy is negative?*

T24 For Coulomb and gravitational force (which change monotonically) with the choice of zero of potential energy at infinite separation, the statement you have made is correct. But it is obvious that your criterion cannot be true in general. The sign or the magnitude of the potential energy at any one point does not have any significance; it can be assigned any desired value, positive or negative, by appropriate choice of the arbitrary additive constant. For example, the harmonic oscillator potential energy $\frac{1}{2}kx^2$ is positive, though the force is always attractive.

The general criterion is: force between two bodies (at a given separation) is attractive if the change ΔV in the potential energy, with an increase in the separation distance between the bodies, is positive.

S25 *How should one decide whether a system will be bound or not for a given energy?*

T25 The total energy of a system (E) is the sum of the potential energy (V) and the kinetic energy (T) of its constituents,

$$E = T + V.$$

V, and hence E, can be positive, negative or zero; the kinetic energy of a system, clearly, must always be non-negative.

† The force on charge 1, for example, has to be calculated as

$$\boldsymbol{F}_1 = -\left(\frac{\partial V}{\partial x_1}\hat{i} + \frac{\partial V}{\partial y_1}\hat{j} + \frac{\partial V}{\partial z_1}\hat{k}\right).$$

Now, the total energy is a constant of motion for an isolated system. The specified value of E, and the form of V, determine the regions in which the kinetic energy of the system can be non-negative. Only these regions are accessible to the system.

Let us denote by $V(\infty)$ the potential energy of a system when the separation between the constituents is infinite.(See Fig. 6.2). For $E > V(\infty)$, evidently, $T > 0$ for infinite separation and the constituents of the system are free to move infinitely apart from each other. Such a system is said to be free.

On the other hand, if $E < V(\infty)$, $T < 0$ for infinite separation which is not allowed. Clearly, in this case, the constituents must always stay within a finite distance of each other. Such a system, whose constituents are confined to a finite region of space, is called a bound system.

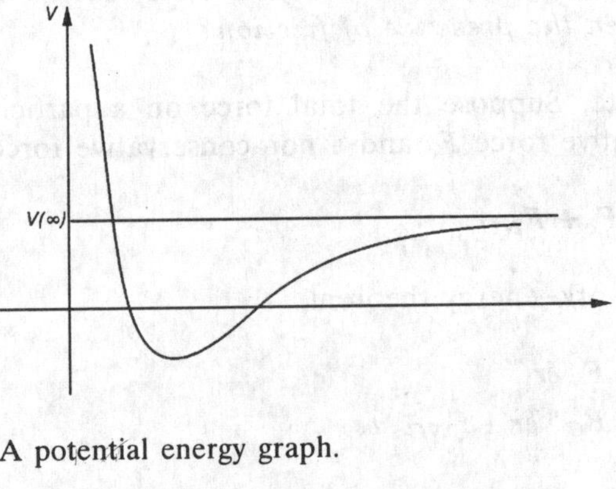

Fig. 6.2 A potential energy graph.

S26 *You have said earlier that friction is not a conservative force. How do you see that?*

T26 It is easy to see that using the criterion of path-independence; in particular, I can check if the work done by friction over a closed path is zero.

Suppose an object is dragged around on a rough surface and brought back to the starting point. No matter which path you follow, the force of kinetic friction opposes the relative motion between the two surfaces at every point and, thus, always acts in a direction opposite to that of the displacement. Consequently

$$\oint \boldsymbol{F}_{\text{friction}} \cdot d\boldsymbol{r} < 0.$$

Suppose, on the other hand, that an object is kept on a trolley and the trolley is moved around. The force responsible for the motion of the object is the force of static friction between the surfaces of the object and the trolley in contact, and this force of friction acts in the same direction as that of the displacement of the object. If the trolley is brought back to the starting point, the object also returns to its initial position. The force of static friction on the object, in this case, acts in the same direction as that of its displacement *throughout the journey* and hence,

$$\oint \boldsymbol{F}_{\text{friction}} \cdot d\boldsymbol{r} > 0.$$

Clearly, in either case, the work done by the force of friction on the object around a closed path is not zero.† Friction, thus, is not a conservative force and there is no such thing as frictional potential energy.

S27 *Since we cannot include the effect of friction in the potential energy function, isn't the definition of total energy $E = T + V$ useless in the presence of friction?*

T27 Not quite. Suppose the total force on a particle is a sum of a conservative force $\boldsymbol{F}_c$ and a non-conservative force $\boldsymbol{F}_{NC}$:

$$\boldsymbol{F} = \boldsymbol{F}_c + \boldsymbol{F}_{NC}.$$

By the work–energy theorem,

$$\begin{aligned} \delta T &= \boldsymbol{F} \cdot \delta \boldsymbol{r} \\ &= \boldsymbol{F}_C \cdot \delta \boldsymbol{r} + \boldsymbol{F}_{NC} \cdot \delta \boldsymbol{r} \end{aligned}$$

Now, since potential energy can be defined for a conservative force, we can write

$$\delta T = -\delta V + \boldsymbol{F}_{NC} \cdot \delta \boldsymbol{r},$$

This, after integration, can be written in the form

$$\Delta E = \Delta(T + V) = \int_1^2 \boldsymbol{F}_{NC} \cdot d\boldsymbol{r}. \tag{7}$$

Thus, the work done by the non-conservative forces equals the change in the total energy.

S28 *It will help if you take some illustrative problems at this point.*

† In the second case, the work done on the trolley by the force of static friction between the body and the trolley is negative. The net work done by the static friction on the assembly is zero.

T28 Suppose a body starts from rest and slides down a rough inclined plane, the coefficient of kinetic friction between the body and the surface being μ. If the distance travelled by the body along the plane is ℓ and if the angle the inclined plane makes with the horizontal is θ (see Fig. 6.3), calculate the speed with which the body reaches the ground.

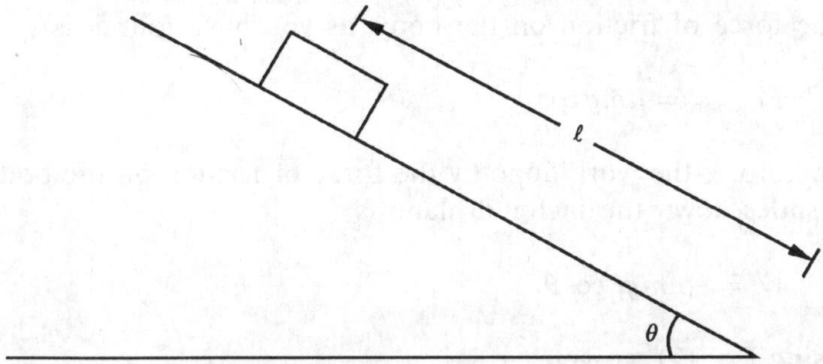

Fig. 6.3 A body sliding down a rough inclined plane.

S29 *This doesn't seem to be a difficult problem. The normal force on the body must be equal to the component of the weight of the body perpendicular to the plane. So* $N = mg\cos\theta$.

The force of friction, therefore, is $F_{\text{friction}} = \mu mg\cos\theta$.

The net force on the body, acting down the inclined plane, is the resultant of the force of friction and the component of the weight down the plane: $mg\sin\theta - \mu mg\cos\theta$.

Hence, the acceleration of the body $a = g\sin\theta - \mu g\cos\theta$.

It is given that the initial velocity is $v_i = 0$ *and the distance travelled is* $s = \ell$. *The kinematical relation* $v_f^2 - v_i^2 = 2as$ *gives the final velocity* $v_f = \sqrt{2g\ell(\sin\theta - \mu\cos\theta)}$.

T29 That is right. But, alternatively, the problem may be solved employing energy considerations. Because of the simultaneous presence of the force of friction and gravity, the form of the work–energy theorem, as stated in T27 is appropriate.

The total energy of the body at the highest point is its potential energy:

$$E_i = mg\ell\sin\theta.$$

At the lowest point all the energy of the body is in the form of kinetic energy:

$$E_f = \tfrac{1}{2}mv_f^2$$

Hence $\Delta E = \tfrac{1}{2}mv_f^2 - mg\ell \sin\theta$.

The force of friction on the body, as you have found, is

$$F_{\text{fricuon}} = \mu mg \cos\theta.$$

Therefore, the work done by the force of friction on the body while it slides down the inclined plane is

$$W = -\mu mg\ell \cos\theta.$$

Using Eq. (7) we get

$$\tfrac{1}{2}mv_f^2 - mg\ell \sin\theta = -\mu mg\ell \cos\theta,$$

which gives the same final velocity.

In this problem the two methods are equally easy. But this is not always the case. Take another example.

Consider a frictionless loop-the-loop apparatus, with radius R, as shown in Fig. 6.4. What is the minimum height H from which the body must be released so that it just completes a loop?

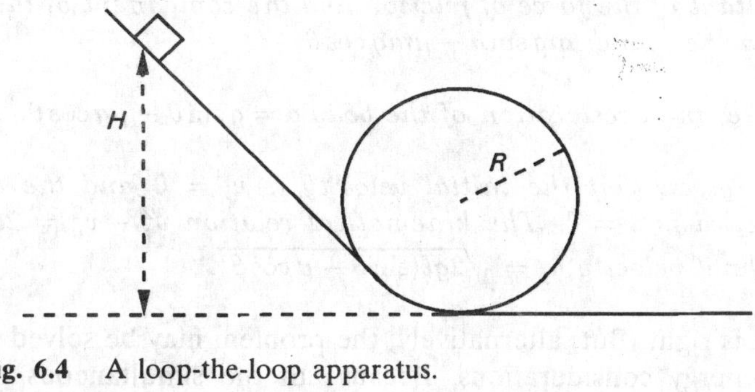

Fig. 6.4 A loop-the-loop apparatus.

S30 *Oh, that is a standard problem. At the highest point of the circular track, in general, the weight of the body and the normal force on the body by the track together provide the necessary centripetal force on the body. When the body just clears the highest point, the normal force on it is zero. Therefore*

$$mg = \frac{mv^2}{R}.$$

The total energy of the body at this point, thus, is

$$\frac{1}{2}mv^2 + mg(2R) = \frac{5}{2}mgR.$$

If the body starts from rest, from a height H, then it must have so much of potential energy at the initial position. Clearly, the height $H = 5R/2$.

T30 That is quite correct. Note you did not solve that the problem of detailed motion of the body along the track using Newton's laws. You solved the problem using only energy considerations. This was possible because the condition for clearing the highest point that you employed depends only on the magnitude of the velocity there. This is what happens in many situations. You can get some interesting features from energy considerations alone even if you cannot solve the full dynamics of a problem.

Conservation of Energy

S31 *Equation (7) seems to imply that the law of conservation of energy is not valid when friction is present. Is that so?*

T31 You must get this clear. Equation (6) expresses conservation of the total energy (kinetic + potential) of a body moving under an _external_ conservative force. The conservation of total energy of an _isolated_ system, on the other hand, is a general principle of physics. It is important to appreciate the difference between the two. In particular, the word 'total' does not mean the same thing in the two statements. The difference is sometimes made clear by calling E of Eq. (6) as the total _mechanical_ energy.

When friction is present, the total mechanical energy of a body is not conserved. This is what is expressed in Eq. (7). But if you consider the system consisting of the body, the sources of the forces (both conservative and non-conservative) and the environment in interaction with the body, then the total energy of the isolated system thus defined is conserved. Note that the total energy here includes not only the mechanical energy but also the various other possible forms of energy such as, internal thermal energy, sound, radiation etc.

For example, the total mechanical energy of a body moving on a rough table is not conserved. However, the body and the table together form an (approximately) isolated system. The total energy of this system, which includes the kinetic energy of the body and the internal (thermal) energy of the body and the table, is (approximately) conserved. We say 'approximately' because we are ignoring, for example, energy lost in the form of sound and other radiations. If everything is included, energy conservation is exact.

S32 *You said that Eq. (6) refers to the conservation of the total mechanical energy of a body under an external conservative force. Why did you emphasize the word 'external' here?*

T32 This is an important point. You see, forces on a body are always exerted by some other bodies. In accordance with Newton's III law, the body exerts equal and opposite forces on the other bodies, and thereby affects their motion. In general, therefore, we must consider the dynamics of the interacting system of bodies as a whole. In that case, you cannot view the motion of any one body as a motion under fixed external agencies of force.

In some situations, however, the agencies, i.e. other bodies causing forces on a body may be practically stationary. This may happen if these other bodies are far more massive than the body under consideration, or if they are constrained by some other means. For example, for the problem of the motion of a stone under earth's gravity, the earth may be regarded as practically stationary. In such situations, it is approximately correct to treat the problem as a single-body problem under a fixed external force. This amounts to a great simplification. Problems of this kind are known in physics as 'external force problems'. Indeed much of our discussion so far implicitly pertains to this situation.

S33 *What is the generalization of the energy conservation equation for a system of interacting particles?*

T33 For an isolated system of particles interacting via conservative forces between pairs of particles, it can be shown using Newton's laws that the sum of the kinetic energy of all the particles and the potential energy between pairs of particles is conserved:

$$\sum_{t=1}^{N} \frac{1}{2} m_i v_i^2 + \frac{1}{2} \sum_{\substack{ij=1 \\ i \neq j}}^{N} V_{ij} = \text{constant},$$

where V_{ij} is the potential energy of mutual interaction between the ith particle and the jth particle. (The sum denoting the potential energy of the system contains pairs of terms such as V_{12} , V_{21} etc. This amounts to double counting. The factor of 1/2 compensates for this.)

However, when all bodies in a system except one, say the kth body, are practically stationary, the total kinetic energy of the system is just the kinetic energy of the kth body. The summation denoting the total kinetic energy of the system, then, reduces to a single term: $\frac{1}{2}m_k v_k^2$. Similarly, the potential energy, which is determined by all the particles of the system, reduces to a function of $\boldsymbol{r}_k$ alone.

Thus, when all particles except one, of a system of particles, are constrained from moving, the problem reduces to that of motion of a single particle under external forces; and the total mechanical energy of the particle is conserved.

S34 *It seems to me that energy is not conserved as we go from one-frame of reference to another. For example, relative to a moving train's frame of reference, all the surroundings: the trees, buildings, even the distant mountains have a kinetic energy of motion. Relative to the ground's frame these same objects are at rest and possess no kinetic energy. What is the source of the huge extra energy of objects in the moving train's frame of reference?*

T34 This is a major confusion among students which you must clear up. First, be clear about what the energy conservation law says. It says that <u>in any given inertial frame of reference</u> the total energy of an isolated system is <u>unchanged in time</u>. The law does <u>*not*</u> claim that the total energy is unchanged from one frame to another in relative motion. That claim would be absurd. Kinetic energy of an object does transform to a different value as we go from one frame to another. This means that the total energy of an isolated system is a constant in time provided that the frame is inertial; but the value of this constant varies from one inertial frame to another.

It is, therefore, meaningless to look for a physical agency or a source which provides the 'huge extra energy' to the mountain in the moving train's frame of reference. That kind of energy arithmetic is to be done in a given frame of reference, not across different frames of reference. The difference in total energy between frames in relative motion has purely kinematic origin.

Another way of putting the same thing is as follows. For the train to come to a certain uniform velocity, it must be accelerated from rest for a certain time. During this time the train's frame is non-inertial, and all objects in the universe experience pseudo-forces relative to this frame. It is the work done by the pseudo-forces that results in the increased kinetic energy of the objects: the trees, the buildings, the mountains, and so on. Of course, the pseudo-forces have no material origin and thus the increase in the kinetic energy, as stated above, is purely kinematic in origin.

S35 *One last query about the law of conservation of energy. Is it an independent principle of physics, or does it follow from the laws of motion?*

T35 As already said, for a mechanical system of particles interacting via conservative forces the principle follows from Newton's laws of motion.

For any system of charged particles and electromagnetic fields, the principle can be seen as a consequence of the laws of electromagnetism and Newton's laws. In this case the total energy comprises of the kinetic energy of the particles and the energy associated with the electric and magnetic fields.

In general, however, the principle is accepted as an independent law of physics (the first law of thermodynamics).

There is another related principle: we believe that under identical conditions an experiment performed today will yield the same results as it did at some time in the past, or as it will yield if performed any time in the future. That is, the laws of physics possess time-displacement invariance. In more advanced theory, the principle of conservation of energy can be shown to be a consequence of this invariance. Of course, the validity of the time-displacement invariance cannot be proved from any first principles; it is an assumption corroborated by our experience.

Conservation of Momentum

S36 *What about the conservation of linear momentum? Does it follow from the laws of motion? Or is it also an independent principle of physics?*

T36 You must never forget to add the qualifier, 'of an isolated system'. The total linear momentum of an isolated system is conserved.

In ordinary mechanics of a system of particles, this principle follows straight from the II and the III laws of motion.

We have seen that, in the relativistic domain, Newton's II law is applicable with a properly modified definition of momentum. Also, Newton's III law is valid in the relativistic domain for contact forces.

Hence, for a system of particles interacting via contact forces, the total linear momentum is conserved in this domain.

Electromagnetic forces are non-contact forces, and in the relativistic description, they are seen as mediated by fields which propagate at the speed of light. The III law is then no longer true as, for example, for forces between moving charges. It turns out that it is possible to define the momentum of an electromagnetic field so that the total momentum of a system of charged particles and fields is conserved. The proof of this statement is beyond our scope here.

In general, however, the principle of momentum conservation must also be accepted as an independent principle of physics.

I said in T35 that the principle of energy conservation is related to the time-displacement invariance. Likewise, the momentum-conservation principle is related to another invariance principle called the space-translation invariance. This means that under identical conditions, the outcome of an experiment will be the same whether it is performed here, there or anywhere else. This amounts to the physically reasonable assumption that space is homogeneous.

S37 *Is the momentum conservation law true in a non-inertial frame of reference?*

T37 Of course not. In a non-inertial frame, an isolated system, i.e. a system without any external material force is still subject to pseudo-forces. Its momentum, therefore, is not conserved. This law is only applicable in an inertial frame.

S38 *For both elastic and inelastic collisions the total linear momentum is conserved; but total kinetic energy is conserved for elastic collisions only. What is behind this?*

T38 During collision between bodies A and B, whatever the nature of forces between them, the force on A by B is the negative of the force on B by A at each instant according to the III law. This is enough to ensure that the total momentum of the two bodies is conserved.

There is no analogous reason that guarantees the conservation of total <u>kinetic</u> energy of the two bodies. No doubt, the total energy of an isolated system is conserved. But as we know, this total energy includes energy in any form: kinetic energy, translational and rotational, potential energy associated with any conservative forces in the system, internal (thermal) energy of bodies etc. In an inelastic collision, depending on the nature of the interaction between the two bodies, part of the initial kinetic energy may get transformed into any of the other possible forms. Consequently the total kinetic energy of bodies need not be conserved in a collision, although their total momentum is always conserved.

S39 *Then what is it that ensures the conservation of total kinetic energy in the case of elastic collisions?*

T39 It is all a question of the nature of the force that acts during a collision. If this force happens to be conservative, the collision is elastic. For example, if during collision the bodies simply get compressed, say like elastic springs, then the force between them is conservative like the usual spring force. This, of course, is not a realistic picture of the force between colliding bodies, but it illustrates the point.

S40 *I am still not clear about the connection between elastic collisions and the conservative nature of force. Please elaborate.*

T40 We have already seen (T19) that, for a conservative force, a potential energy can be defined such that the sum of kinetic energy and potential energy of the system is conserved. Now, the force associated with collision has one obvious property: it is non-zero only for such separation between the bodies where they can be in contact. Thus, for the collision of two billiard balls, the force is zero when the separation between them (i.e. the distance between their centres) exceeds the sum of their radii. In this entire region the potential energy is a constant, which we can choose to be zero (see T22).

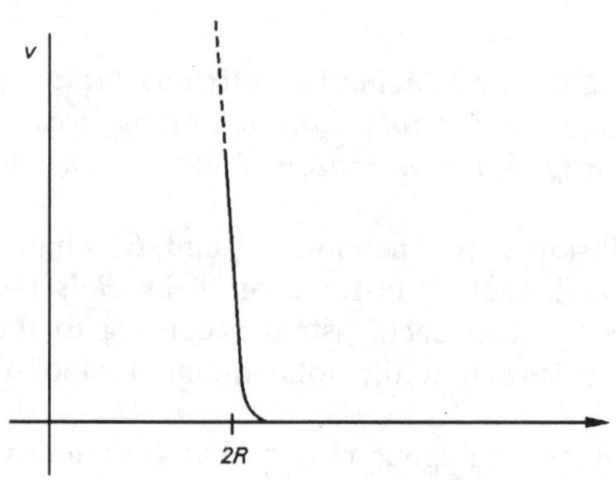

Fig. 6.5 A schematic graph of the potential energy of interaction between two balls, each of radius R, as a function of separation between their centres.

Figure 6.5 shows a possible graph for potential energy versus separation, corresponding to a collision of two balls each of radius R.

Note $V(r)$ and the force between the balls both are zero for $r > 2R$ and shoot up for $r < 2R$, showing the sharply increasing repulsive force as the balls get compressed further and further. (Actually, the graph is sensible only for values of r near $2R$. For smaller r, the bodies are compressed beyond their elastic limit; and the notion of potential energy is inapplicable.)

Thus for a conservative force which is zero at large distances, the initial total energy is simply the initial total kinetic energy. Likewise, the final total energy equals the final total kinetic energy. Since the sum of kinetic and potential energy is a constant throughout for any conservative force, the total initial kinetic energy of the system must equal the total final kinetic energy.

S41 *What happens during the short time of collision when the bodies are in contact? Is the total momentum conserved during this time? Is the total kinetic energy conserved during this time, if the collision is elastic?*

T41 That should be clear to you from what we have already discussed. At each instant during the collision, whatever be the type of collision (elastic or inelastic), the III law is always valid. Since the mutual forces are equal and opposite, the change in momentum of body A is the negative of the change in momentum of body B. The total momentum of bodies A and B, therefore, remains the same at every instant before, during and after the collision.

If the collision is elastic, the initial (i.e. when the bodies are apart before collision) total kinetic energy is equal to the final (i.e. when the bodies move apart again after collision) total kinetic energy. <u>During</u> the collision, however, the total kinetic energy is not conserved even if the collision is elastic. The kinetic energy decreases initially during an elastic collision, being progressively transformed into potential energy. The bodies may come to a rest momentarily if the collision is one-dimensional. During the later part of the collision, the potential energy gets transformed back to kinetic energy. The sum of the total kinetic energy and the potential energy is, of course, constant at every instant for an elastic collision.

S42 *Are the laws of conservation of momentum and energy independent of each other? I mean, in a process, if one condition is satisfied, will the other be satisfied automatically, or is that a separate requirement?*

T42 These are two independent conservation laws. One does not follow from the other. They impose separate requirements on a process. For a process to occur, the conditions of energy and momentum conservation must be simultaneously satisfied. This greatly restricts possible processes in nature.

Figure 6.6 gives three outcomes of a one-dimensional collision. All of them satisfy the condition of momentum conservation. Yet if we are given that the collision is elastic, only the process in Fig. 6.6(a) is possible since that alone conserves total kinetic energy. The remaining two processes are not possible outcomes of an elastic collision.

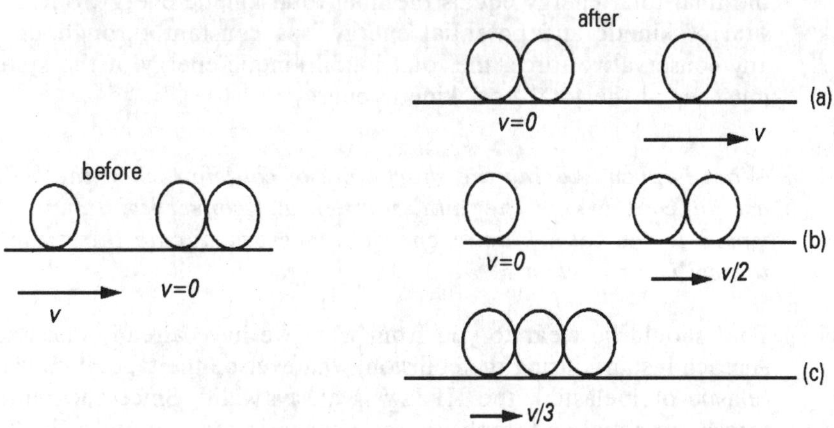

Fig. 6.6 Outcomes of a one-dimensional collision. The only outcome allowed by energy-momentum conservation is the one in (a).

S43 *Any more examples of the application of energy–momentum conservation?*

T43 This is something that one keeps at the back of one's mind all the time. Indeed, in the atomic and sub-atomic domain, energy–momentum conservation is one of the first things to check in order to rule out a process. For example, consider a process in which a free proton gets converted into a neutron by emitting a pion

$$p \to n + \pi^+.$$

It is forbidden by energy–momentum conservation, although it satisfies many other requirements such as charge conservation, baryon number conservation† etc.

† Particles capable of experiencing the nuclear strong force are called 'hadrons'. They are classified in two categories: hadrons with integral spin, such as pions, are called mesons; and hadrons with half odd integral spin, such as protons and neutrons, are called baryons. Each particle has an associated baryon number B. Mesons have $B = 0$, baryons have $B = 1$ and antibaryons have $B = -1$. Baryon number, like charge, is an additive quantum number. Total baryon number is conserved in every process of nature observed so far.

S44 *How do you see that?*

T44 Suppose the proton is initially at rest. Then conservation of momentum, then requires

$$\boldsymbol{p}_n = -\boldsymbol{p}_{\pi^+}$$

and conservation of total energy requires

$$m_p c^2 = \sqrt{p_n^2 c^2 + m_n^2 c^4} + \sqrt{p_{\pi^+}^2 c^2 + m_{\pi^+}^2 c^4}.$$

It is clearly impossible to satisfy the above equation since $m_p < m_n$.

S45 *But what if I give enough kinetic energy to the proton so that its total energy is greater than the sum of the rest-mass energy of n and π^+? Relativity allows energy–mass conversion, so a sufficiently energetic proton should be able to decay into a neutron and a pi-meson.*

T45 That is a serious error. You have made this mistake because you are looking at energy conservation in isolation from momentum conservation. As already stressed, a process must satisfy both the conservation laws at the same time.

If initially the proton has momentum $\boldsymbol{p}_p$ then energy–momentum conservation conditions are:

$$\boldsymbol{p}_p = \boldsymbol{p}_n + \boldsymbol{p}_{\pi^+};$$

$$\sqrt{p_p^2 c^2 + m_p^2 c^4} = \sqrt{p_n^2 c^2 + m_n^2 c^4} + \sqrt{p_{\pi^+}^2 c^2 + m_{\pi^+}^2 c^4}.$$

It can be shown algebraically that the two conditions are impossible to satisfy simultaneously. That is, the process is forbidden.

One could arrive at the same conclusion by a more physical argument. Let us consider the situation relative to a frame that is moving with the same velocity as the proton. In this frame, the proton is at rest. By the same reasoning as in T44, the process is not allowed relative to this frame. Now, the same process cannot be forbidden relative to one frame and permitted relative to another. Thus, the decay of a high energy proton into a neutron and a π^+ meson is not possible.

S46 *The decay $A \to B + C$ is not allowed if $m_A < m_B + m_C$. But the reaction $A + B \to C + D$ is allowed even if $m_A + m_B < m_C + m_D$. Why?*

T46 It is all a question of being able to satisfy simultaneously both energy and momentum conservation. We have seen this is not possible in the first case. But for the second case it is possible, provided the reactants are given enough energy.

There is another important thing to note in case of two-body decay: $A \to B + C$. For the decay of particle A at rest, the decay products must always come out with definite energy and momentum of definite magnitude, as fixed by equations like those in T44.

S47 *Yes, I see that. But what is so significant about it?*

T47 Suppose in the decay of a certain particle at rest, seemingly, there are only two products. However, rather than having definite energies, the two decay products are experimentally observed to come out with a spectrum of energies. This experimental finding is incompatible with a two-body decay, and indicates that we are failing to detect some other particle(s)

This is what happened historically with β decay. The β decay of a nucleus apparently gave only two products: the daughter nucleus and an electron. But the electrons did not have a fixed energy; their energy spectrum was observed to be continuous. Pauli reconciled this finding with the energy–momentum conservation by suggesting that, in β decay, the parent nucleus changes to the daughter nucleus by emitting, not just an electron, but an electron and some other particle.† The existence of this particle is now established beyond doubt. It is called the antineutrino.

S48 *Let us return to a classical problem. Consider the motion of a ball on a frictionless floor. If it hits a wall, it rebounces back with some speed. The wall is stationary throughout, and hence the change in its momentum is zero. The momentum of the ball has changed sign. Clearly, the total momentum of the isolated (wall + ball) system has changed. How can that be?*

T48 First of all be clear about the isolated system. The wall is rigidly (almost) connected to the ground, so that by 'wall' we should mean really the whole earth. Thus what we are talking of is the isolated ball + earth system.††

There is no violation of the law of momentum conservation

† Some additional arguments were also advanced for this hypothesis, which need not concern us here.

†† Strictly, this system is not isolated; it is moving under the sun's gravity. Also the earth's frame is not strictly inertial. However, for the problem at hand these effects are negligible.

here. Remember this law is so universally true that in any argument showing its apparent violation, one must detect some error. The error in your argument is in thinking that the wall remains strictly stationary. Let us see this better.

During collision the wall exerts a force on the ball. By the III law, the ball exerts an equal and opposite force on the wall. This causes the wall to recoil back. However, since the wall is much more massive than the ball, its recoil velocity is negligible compared to the ball. That is what makes you think it is at rest.

S49 *Qualitatively this seems alright; but can we not see it more precisely?*

T49 That is quite straightforward to see. Let us denote the mass of the ball by m and that of the wall by M. Let the velocity of the ball before collision be v_b, and let the velocities of the ball and the wall after collision be v_b' and v_w' respectively. Momentum conservation requires

$$mv_b = mv_b' + Mv_w',$$

and energy conservation, assuming the collision to be elastic, gives

$$\frac{1}{2}mv_b^2 = \frac{1}{2}mv_b'^2 + \frac{1}{2}Mv_w'^2.$$

These equations can be easily solved to obtain the velocities after collision as

$$v_b' = \frac{m-M}{m+M}v_b; \quad v_w' = \frac{2m}{m+M}v_b.$$

Evidently, as $M/m \to \infty$, $v_w' \to 0$ and $v_b' \to -v_b$; i.e. for M very large compared to m, v_w' is negligible compared to v_b, and the ball bounces back with the same peed. The momenta of the two, after collision, are given by

$$p_b' = \frac{m-M}{m+M}p_b; \quad p_w' = \frac{2M}{m+M}p_b.$$

Notice, although $v_w' \to 0$ as $M/m \to \infty$, $p_w' \to 2p_b$ in the same limit, i.e. the momentum of the wall is <u>not</u> negligible compared to the momentum of the ball (in fact, it is twice as much here). It is the velocity that is negligible. (The negligible velocity times the very large mass becomes a non-negligible momentum.)

S50 *For the elastic collision that one often considers, the ball is taken to rebounce with the same speed as the initial speed. If, as you say, the wall recoils with some momentum, it should carry away some kinetic energy. The ball must. therefore, rebounce with reduced energy. Is that not so?*

T50 You should appreciate an important point: when a massive body recoils, it may carry non-negligible momentum and yet its recoil energy can be negligible. In the collision between the ball and the wall, the wall shares the momentum but not energy.

S51 *How is that possible?*

T51 From the expressions for v_w' above, we can write the expression for the kinetic energy of the wall, T_w', after collision as

$$T_w' = \frac{4mM}{(m+M)^2} T_b,$$

where T_b is the kinetic energy of the ball before collision. It is clear that $T_w' \to 0$ as $M/m \to \infty$.

A simpler way is to express kinetic energy in terms of momentum:

$$T_b' = \frac{p_b'^2}{2m}; \quad T_w' = \frac{p_w'^2}{2M}.$$

Clearly, since p_b' and p_w' are comparable, $T_w' << T_b'$ if $M >> m$.

S52 *This seems a rather interesting point. Do you come across such situations elsewhere in physics?*

T52 Yes, at many places. For example, when an atom in an energy state E_1 decays to another energy state E_2, a photon of frequency ν is emitted. Do you know how ν is related to the energies E_1 and E_2?

S53 *I know that. It is given in the Bohr model by $E_1 - E_2 = h\nu$.*

T53 You are almost right, but you have ignored the small recoil effect. When an atom at rest decays, the momentum of the emitted photon $h\nu/c$ must be equal and opposite to the recoil momentum of the atom. So the recoil energy of the atom is given by

$$\frac{p^2}{2m} = \frac{h^2\nu^2}{2mc^2}$$

and hence, the photon energy is given by the equation

$$E_1 - E_2 = h\nu + \frac{h^2\nu^2}{2mc^2}.$$

The recoil energy is negligible as compared to the photon energy since $h\nu$ is of the order of a few $e\nu$ whereas mc^2 for an atom is of the order of BeV.

Similar considerations are relevant when a nucleus decays from an excited level E_1 to a lower level E_2, emitting a gamma ray. In this case the energy of the emitted photon is much larger (of the order of Mev) and, therefore, the recoil momentum and energy of the nucleus are much greater. There is a way, however, of preventing the nucleus from recoiling. If the nucleus is embedded in a crystal, then, roughly speaking, the entire crystal recoils. Because of its large effective mass, the recoiling system carries little energy, as in the ball + wall problem, while it carries away as much momentum as required by momentum conservation. This is the idea underlying a well-known method of high-resolution nuclear spectroscopy: Mossabauer spectroscopy.

S54 *Let me return to our discussion in T43. According to that a process like $p \rightarrow n + e^+ + v$ is not allowed by energy–momentum conservation since the proton mass is less than the neutron mass. Yet this is the process that gives rise to β^+ emission. How do you understand that?*

T54 You must be careful here. Energy–momentum conservation says that a <u>free</u> proton cannot decay as above. If the proton is bound, the decay is possible.

S55 *I did not get you. How does the proton being 'bound' help?*

T55 In this case, the decay to be considered is:

(p + rest of the parent nucleus) $\rightarrow$ (n + rest of the daughter nucleus) + $e^+ + v$.

The important thing is that the rest of the nucleus is not isolated from the process; it shares energy and momentum as appropriate for conservation. The only condition is that the mass of the parent nucleus should be greater than the sum of the masses of the products. The binding energy of the proton with the rest of the parent nucleus is not the same as the binding energy of the neutron with the rest of the daughter nucleus. Indeed you can say that the difference in these binding energies is what provides the necessary extra energy that the proton needs in order to decay into a more massive neutron.

S56 *This looks like an important point. Is there any other example?*

T56 We come across a related, but not completely similar situation in photoelectric emission. A process

$$e^-_{\text{free}} + \gamma \to e^-$$

is not allowed by energy-momentum conservation. (To see this, observe the process relative to the final electron's rest frame. The initial electron and the photon must have equal and opposite momenta in this frame. Consequently, energy conservation is violated in the process.)

S57 *But photoelectric emission is due to the absorption of a photon from incident light by an electron in a metal. How does it take place at all if the above process is not allowed?*

T57 The answer is that the initial electron is not free; it is bound in the metal. The following process is not forbidden by energy–momentum conservation:

$$\text{lattice} + e^-_{\text{bound}} + \gamma \to \text{lattice} + e^-_{\text{free}}.$$

At this point you must note one thing. Though the lattice shares momentum in the process, its share of energy is completely negligible. The reason this is so is exactly as in the case of the ball + wall problem. The mass of the lattice is so large that $p^2_{\text{lattice}}/2M$ is nearly zero.

S58 *One final question. A nuclear reaction is always cited as an example of mass–energy interconversion. Why should this be so only for a nuclear reaction? Why can't mass be converted into energy or vice versa in a chemical reaction?*

T58 This is a common misconception. Actually, as far as mass–energy interconversion is concerned, there is no difference <u>in principle</u> between a nuclear reaction and a chemical reaction; the difference is only in the magnitudes of the energies involved. Mass gets converted into energy (or vice versa) even in a chemical reaction.

S59 *I am surprised. Please elaborate.*

T59 Let us see how the energetics of a chemical reaction was looked at before the advent of the ideas of mass–energy interconversion. Take, for example, the reaction:

$$NaOH + HCl \rightarrow NaCl + H_2O.$$

This is what was believed :

1) A chemical reaction is simply a rearrangement of atoms in different combinations. No atoms are created or destroyed in a chemical reaction. (This is the reason we need to balance a chemical equation.)

2) Atoms are bound with different binding energies in different chemical compounds. The total binding energy in the reactants may be different from the total binding energy in the products. This difference in total binding energies on the two sides of a reaction appears as heat produced or absorbed. In an exothermic reaction as above, the binding energy on the right ($B.E._{r.h.s.}$) is greater than the binding energy on the left ($B.E._{l.h.s.}$). Thus,

$B.E._{r.h.s.} - B.E._{l.h.s.} = Q$, the net heat produced.

3) Mass of a compound is a sum of the masses of its constituent atoms; e.g., mass of NaCl = mass of Na+ mass of Cl; and similarly for other compounds. Since there is no creation or destruction of individual atoms, the total mass of the reactants is equal to the total mass of the products. This is the principle of conservation of mass in a chemical reaction.

S60 *What is wrong with this view-point?*

T60 The first two statements continue to be correct. Statement (3), however, is not strictly correct from a relativistic view-point.

The basic thing you must know is that, in relativistic mechanics, the rest mass of a bound composite system is not equal to the sum of the rest masses of its constituent parts. By definition, binding energy (B.E.) is the energy required to unbond the parts of a system. By mass–energy equivalence, this implies for a bound system such as NaCl

$$m_{NaCl}c^2 + B.E. = m_{Na}c^2 + m_{Cl}c^2,$$

where m_x is the mass of x. This means

$$m_{NaCl} = m_{Na} + m_{Cl} - \delta m; \quad \delta\ m = B.E./c^2.$$

δm is called 'mass defect'.

Now, since the total binding energy is not the same on the two sides of the reaction, it is clear from the above equation that the total mass defects on the two sides are unequal. This implies, in turn, that the total mass on the left side (reactants) is not the same as the total mass of the right side (products). Thus mass is not conserved in a chemical reaction.

In short, from a relativistic view-point, Statement (2) above (T59) is completely equivalent to the statement that the heat produced in a chemical reaction is due to the conversion of part of the mass of the reactants into energy.

S61 *If that is so, why does everybody cite nuclear reactions as example of mass–energy interconversion?*

T61 It is simply a matter of magnitudes. In chemical compounds the binding energies are of the order of eV. The nucleonic binding energy in nuclei, on the other hand, is typically a few MeV. Consequently, the mass defects in nuclei are much larger than those in chemical compounds. In fact the mass defect of a chemical compound is hardly measurable. In nuclei, mass defects are appreciable. Therefore mass–energy interconversion is usually invoked in nuclear reactions and not in chemical reactions.

7

Systems of Particles

Laws of Motion for Many-Particle Systems

S1 *So far we have only dealt with single particle mechanics. How does one handle systems of many particles?*

T1 We apply Newton's II law to each particle of the system:

$$\frac{d\boldsymbol{p}_i}{dt} = \boldsymbol{F}_i.$$

The force $\boldsymbol{F}_i$ on the ith particle can be expressed as a sum of an external force $\boldsymbol{F}_i^{\text{ext}}$ and an internal force $\boldsymbol{F}_i^{\text{int}}$ due to all the other particles in the system:

$$\begin{aligned}\boldsymbol{F}_i &= \boldsymbol{F}_i^{\text{ext}} + \boldsymbol{F}_i^{\text{int}} \\ &= \boldsymbol{F}_i^{\text{ext}} + \sum_{\substack{j=1 \\ j \neq i}}^{N} \boldsymbol{F}_{ij}\text{,}\end{aligned}$$

where N is the number of particles in the system. The mutual forces between any pair of particles satisfy the III law:

$$\boldsymbol{F}_{ij} = -\boldsymbol{F}_{ji}.$$

The total linear momentum of the system is

$$\boldsymbol{P} = \sum_{i=1}^{N} \boldsymbol{p}_i.$$

Consequently,

$$\frac{d\boldsymbol{P}}{dt} = \sum_{i=1}^{N} \frac{d\boldsymbol{p}_i}{dt} = \sum_{t=i}^{N} \boldsymbol{F}_i^{\text{ext}} + \sum_{\substack{ij=1 \\ i \neq j}}^{N} \boldsymbol{F}_{ij}.$$

The second term contains pairs of terms like $\boldsymbol{F}_{34} + \boldsymbol{F}_{43}, \boldsymbol{F}_{15} + \boldsymbol{F}_{51}$

etc. which cancel out due to III law. The first term is the net external force on the system. We therefore write

$$\frac{d\boldsymbol{P}}{dt} = \boldsymbol{F}^{\text{ext}} \tag{1}$$

This equation looks like the II law of motion, but remember the III law has been used in arriving at it.

S2 *One often uses the law $d\boldsymbol{L}/dt = \boldsymbol{N}$. Is this an independent law of motion?*

T2 For a single particle it is a direct consequence of Newton's II law. Since $d\boldsymbol{p}/dt = \boldsymbol{F}$, we have $\boldsymbol{r} \times d\boldsymbol{p}/dt = \boldsymbol{r} \times \boldsymbol{F}$, i.e.

$$\frac{d\boldsymbol{L}}{dt} = \boldsymbol{N}.$$

For a system of particles, the total angular momentum is

$$\boldsymbol{L} = \sum_{i=1}^{N} \boldsymbol{r}_i \times \boldsymbol{p}_i.$$

Therefore

$$\begin{aligned}\frac{d\boldsymbol{L}}{dt} &= \frac{d}{dt}\sum_{i=1}^{N} \boldsymbol{r}_i \times \boldsymbol{p}_i = \sum_{i=1}^{N} \boldsymbol{r}_i \times \frac{d\boldsymbol{p}_i}{dt} \\ &= \sum_{i=1}^{N} \boldsymbol{r}_i \times \boldsymbol{F}_i^{\text{ext}} + \sum_{\substack{ij=1 \\ i\neq j}}^{N} \boldsymbol{r}_i \times \boldsymbol{F}_{ij}\end{aligned}$$

The last term contains pairs of terms like $\boldsymbol{r}_\ell \times \boldsymbol{F}_{\ell m} + \boldsymbol{r}_{\text{m}} \times \boldsymbol{F}_{m\ell}$. Since by the III law, $\boldsymbol{F}_{\ell m} = -\boldsymbol{F}_{m\ell}$, we may write the pair as $(\boldsymbol{r}_\ell - \boldsymbol{r}_m) \times \boldsymbol{F}_{\ell m}$. Now, if the mutual forces between a pair of particles, in addition to being equal and opposite, act along the line joining the particles, i.e. if $\boldsymbol{r}_\ell - \boldsymbol{r}_m$ and $\boldsymbol{F}_{\ell m}$ are parallel, their cross-product vanishes. It is then possible to write

$$\frac{d\boldsymbol{L}}{dt} = \boldsymbol{N}^{\text{ext}},$$

where

$$\boldsymbol{N}^{\text{ext}} = \sum_{t=1}^{N} \boldsymbol{r}_i \times \boldsymbol{F}_i^{\text{ext}}$$

is the total external torque on the system.

Note again that, for arriving at this result, we have used the III law in addition to the II law of motion. Also note that in T1 it is the weak form of the III law that is used; in the present case we need the strong form of the III law.

S3 *So it is true that the equation $d\boldsymbol{L}/dt = \boldsymbol{N}$ is merely a consequence of the laws of motion; it is not an independent law.*

T3 Yes, but don't be misled by that. For a single particle, the equation $d\boldsymbol{L}/dt = \boldsymbol{N}$ follows from the II law $d\boldsymbol{p}/dt = \boldsymbol{F}$; the two are not independent equations.

For a system of particles we have the equations $d\boldsymbol{L}/dt = \boldsymbol{N}^{\text{ext}}$ and $d\boldsymbol{P}/dt = \boldsymbol{F}^{\text{ext}}$ which are similar in appearance to the corresponding single particle equations. They make use of both the II and the III law. But they are independent equations; one does not follow from the other.

The usual laws of conservation of linear momentum and angular momentum of an isolated system follow immediately. If $\boldsymbol{F}^{\text{ext}} = 0$, $\boldsymbol{P}$ is conserved. If $\boldsymbol{N}^{\text{ext}} = 0$, $\boldsymbol{L}$ is conserved. For an isolated system $\boldsymbol{F}^{\text{ext}} = \boldsymbol{N}^{\text{ext}} = 0$; so both $\boldsymbol{P}$ and $\boldsymbol{L}$ are conserved.

S4 *Is it possible to have $\boldsymbol{F}^{\text{ext}} = 0$ but $\boldsymbol{N}^{\text{ext}}$ non-zero or vice-versa?*

T4 Of course it is possible, and you must be familiar with such situations. A simple example is a system acted upon by an external couple. A couple consists of two equal and opposite forces with parallel, but not coincident, lines of action. Clearly $\boldsymbol{F}^{\text{ext}} = 0$ in this case, but $\boldsymbol{N}^{\text{ext}} \neq 0$.

A number of concurrent forces acting on a system may give rise to a net force so that $\boldsymbol{F}^{\text{ext}} \neq 0$. But if moments are calculated about the common point of action, $\boldsymbol{N}^{\text{ext}} = 0$.

S5 *What if torques are calculated about some other point?*

T5 Torque, or the moment of a force, depends on the point about which it is calculated. So in the above example, if moments are calculated about some other point, $\boldsymbol{N}^{\text{ext}}$ may be non-zero. Thus in general, a statement about torques always refers to some point.

When, however, the total force on a system is zero, the total torque on the system is independent of the point about which it is calculated. The result can be easily proved. If $\boldsymbol{r}_O$ is the position vector of a point O' relative to O, $\boldsymbol{r}_i = \boldsymbol{r}_O + \boldsymbol{r}_i'$ where $\boldsymbol{r}_i$ and $\boldsymbol{r}_i'$ are the position vectors of the point of application of $\boldsymbol{F}_i$ relative to points O and O' respectively. Clearly,

$$\sum_{i=1}^{N} \boldsymbol{r}_i \times \boldsymbol{F}_i = \boldsymbol{r}_0 \times \sum_{i=1}^{N} \boldsymbol{F}_i + \sum_{i=1}^{N} \boldsymbol{r}_i{}' \times \boldsymbol{F}_i$$

$$= \sum_{i=1}^{N} \boldsymbol{r}_i{}' \times \boldsymbol{F}_i$$

$$\text{if} \qquad \sum \boldsymbol{F}_i = 0.$$

A familiar example of such a system is that of a couple. The moment of a couple is independent of the origin. The conditions of static equilibrium of a rigid body are $\sum \boldsymbol{F}_i^{\text{ext}} = 0, \sum \boldsymbol{N}_i^{\text{ext}} = 0$. Here, in view of the above result, the total torque may be calculated about any point.

Centre of Mass

S6 *One so often invokes the centre of mass for a system of particles. What is its significance?*

T6 The location of the centre of mass is defined to be the weighted average of the locations of all the particles:

$$\boldsymbol{R} = \frac{\sum_{i=1}^{N} m_i \boldsymbol{r}_i}{\sum_{i=1}^{N} m_i} = \frac{\sum_{i=1}^{N} m_i \boldsymbol{r}_i}{M}.$$

So in a sense the CM represents the overall location of the system. Note, that this location may or may not be within the matter distribution of the system. For example, the CM of a ring or a hollow sphere is outside, at the centre.

The significance of the CM is the following: many physical quantities of the system neatly separate into two parts, one referring to the motion of the CM and the other referring to motion relative to the CM. Also, several dynamical equations in the CM frame have forms similar to those in the laboratory (inertial) frame, even when the CM frame may be accelerating.

S7 *I did not quite get you.*

T7 Let us start with the linear momentum of a system:

$$\boldsymbol{P} = \sum_{i=1}^{N} m_i \frac{d\boldsymbol{r}_i}{dt} = \frac{d}{dt} \sum_{i=1}^{N} m_i \boldsymbol{r}_i = M \frac{d\boldsymbol{R}}{dt} = M\boldsymbol{V}.$$

Thus the total linear momentum of the system is equal to the mass of the system times the velocity of the CM—the linear momentum of the CM so to speak. So, as far as linear momentum is concerned, the total mass of the system can be considered to be concentrated at the CM. Relative to the CM, the total linear momentum of the system is zero.

S8 *How about the total kinetic energy of the system? Is it equal to kinetic energy of the CM, i.e.* $(1/2)MV^2$*?*

T8 No. That is a possible pitfall. Although the total momentum of a system equals the momentum of the CM, the total kinetic energy of the system is not equal to that of the CM. Let us see this:

Let $\boldsymbol{R}$ be the position vector of the CM relative to the laboratory frame. The relation between $\boldsymbol{r}_i$ (the position vector of the ith particle relative to the laboratory frame) and $\boldsymbol{r}_i'$ (the position vector of the ith particle relative to the CM) is

$$\boldsymbol{r}_i = \boldsymbol{r}_i' + \boldsymbol{R}. \tag{2}$$

Clearly,

$$\boldsymbol{v}_i = \boldsymbol{v}_i' + \boldsymbol{V}. \tag{3}$$

The kinetic energy of the system in the laboratory frame is

$$T = \frac{1}{2} \sum_{i=1}^{N} m_i v_i^2.$$

Substituting for $\boldsymbol{v}_i$ from above we obtain

$$T = \frac{1}{2} M V^2 + \frac{1}{2} \sum_{i=1}^{N} m_i {v_i'}^2 + \boldsymbol{V} \cdot \sum_{i=1}^{N} m_i \boldsymbol{v}_i'.$$

The last term is zero since $\sum m_i \boldsymbol{v}_i' = 0$ in the CM frame. Therefore we get

$$T = \frac{1}{2} M V^2 + \frac{1}{2} \sum_{i=1}^{N} m_i {v_i'}^2.$$

The second term has a simple interpretation. Consider a frame whose origin is moving with the CM velocity $\boldsymbol{V}$ and whose axes have fixed orientation relative to the axes of the laboratory (inertial) frame. We call this the CM frame of the system. We shall choose its origin at the location of the CM. Note, the CM frame so defined has translational motion but no rotational motion relative to the laboratory frame. $\boldsymbol{v}_i'$ then, is the velocity of the ith particle relative to the CM frame. Thus the kinetic energy of the system gets decomposed into two parts:

$$T = T(CM) + T', \tag{4}$$

where $T(CM) = (1/2)MV^2$ is the kinetic energy of the CM in the laboratory frame and $T' = (1/2)\sum m_i v_i'^2$ is the kinetic energy of the system relative to the CM frame.

S9 *How is it that the momentum of the system in the CM frame is equal to zero but the kinetic energy of the system in the same frame is not?*

T9 Zero net momentum in the CM frame does not mean that the particles are stationary relative to that frame; only their momenta add up vectorially to zero. Since the kinetic energy is a non-negative scalar (involving square of velocity), it does not add up to zero.

S10 *What about the angular momentum of the system and the torque acting on it?*

T10 Here again, like kinetic energy, there is a neat decomposition using the CM frame.

We have $$\boldsymbol{L} = \sum_{i=1}^{N} \boldsymbol{r}_i \times \boldsymbol{p}_i = \sum_{i=1}^{N} m_i \boldsymbol{r}_i \times \boldsymbol{v}_i$$

Substituting for $\boldsymbol{r}_i$ and $\boldsymbol{v}_i$ from Eqs. (2) and (3) one gets

$$\boldsymbol{L} = \sum_{i=1}^{N} \boldsymbol{R} \times m_i \boldsymbol{V} + \sum_{i=1}^{N} \boldsymbol{r}_i' \times m_i \boldsymbol{v}_i'$$

$$+ \left(\sum_{i=1}^{N} m_i \boldsymbol{r}_i'\right) \times \boldsymbol{V} + \boldsymbol{R} \times \frac{d}{dt} \sum_{i=1}^{N} m_i \boldsymbol{r}_i'.$$

The last two terms contain $\sum m_i \boldsymbol{r}_i'$ which is zero. Hence, we get

$$\boldsymbol{L} = \boldsymbol{R} \times M\boldsymbol{V} + \sum_{i=1}^{N} \boldsymbol{r}_i' \times \boldsymbol{p}_i'.$$

The first term is the angular momentum of the CM motion about the origin of the laboratory frame and the second term is the angular momentum of the system about the CM in the CM frame. Thus we can write

$$\boldsymbol{L}_O = \boldsymbol{L}_O(CM) + \boldsymbol{L}_{CM}',$$

where $\boldsymbol{L}_O$ is the angular momentum of a system of particles about any point O in the laboratory frame, $\boldsymbol{L}_O$(CM) is the angular momentum of the CM about O in the same frame, and $\boldsymbol{L}_{\text{CM}}'$ is the angular momentum of the system in the CM frame about the CM.

Likewise, the torque $\boldsymbol{N}_O$ around a point O in the laboratory frame can be decomposed as

$$\boldsymbol{N}_O = \sum_{i=1}^{N} \boldsymbol{r}_i \times \boldsymbol{F}_i = \sum_{i=1}^{N} \boldsymbol{r}_i' \times \boldsymbol{F}_i + \boldsymbol{R} \times \sum_{i=1}^{N} \boldsymbol{F}_i.$$

The first term on the right-hand side can be written as

$$\sum_{i=1}^{N} \boldsymbol{r}_i' \times \sum_{j \neq i} \boldsymbol{F}_{ij} + \sum_{i=1}^{N} \boldsymbol{r}_i' \times \boldsymbol{F}_i^{\text{ext}}$$

$$= \sum_{i=1}^{N} \boldsymbol{r}_i' \times \boldsymbol{F}_i^{\text{ext}}$$

since, by the strong form of the III law,

$$\sum_{\substack{i,j=1 \\ j \neq i}}^{N} \boldsymbol{r}_i' \times \boldsymbol{F}_{ij} = 0.$$

Hence, $$\boldsymbol{N}_O = \sum_{i=1}^{N} \boldsymbol{r}_i' \times \boldsymbol{F}_i^{\text{ext}} + \boldsymbol{R} \times \sum_{i=1}^{N} \boldsymbol{F}_i.$$

Now, the internal forces cancel pair-wise so that $\sum \boldsymbol{F}_i = \sum \boldsymbol{F}_i^{\text{ext}} = \boldsymbol{F}^{\text{ext}}$. Therefore

$$\boldsymbol{N}_O = \boldsymbol{R} \times \boldsymbol{F}^{\text{ext}} + \boldsymbol{N}_{\text{CM}}'^{\text{ext}},$$

where $N_{CM}^{\prime ext}$ is the total torque due to the external forces about the CM and $\boldsymbol{R} \times \boldsymbol{F}^{\text{ext}}$ is the torque about O of the total external force acting at the CM. Note the important point that the torques due to internal forces cancel in pairs (assuming the strong form of the III law); thus $N_0 = N_0^{\text{ext}}$ and $N_{\text{CM}}^{\prime} = N_{\text{CM}}^{\prime\,\text{ext}}$.

We thus see that important physical quantities like $\boldsymbol{P}$, $\boldsymbol{T}$, $\boldsymbol{L}$, $\boldsymbol{N}$ have a simple decomposition into two parts: one pertaining to the motion of the CM and the other pertaining to the motion relative to the CM. Note that such a decomposition is not possible relative to just any frame moving with respect to the laboratory frame.† That is what makes the CM frame a very special and useful frame for a system of particles.

S11 *In what way does this decomposition help?*

T11 Take for example the law $d\boldsymbol{L}_0/dt = \boldsymbol{N}_0^{\text{ext}}$ in the laboratory frame. Since both $\boldsymbol{L}_0$ and $\boldsymbol{N}_0^{\text{ext}}$ split as seen above, we have

$$\frac{d\boldsymbol{L}_{CM}^{\prime}}{dt} + \frac{d\boldsymbol{L}_0(CM)}{dt} = \boldsymbol{N}_{CM}^{\prime\,ext} + \boldsymbol{R} \times \boldsymbol{F}^{ext}. \tag{5}$$

Now, $$\frac{d\boldsymbol{L}_0(CM)}{dt} = \frac{d}{dt}(\boldsymbol{R} \times \boldsymbol{P}) = \frac{d\boldsymbol{R}}{dt} \times \boldsymbol{P} + \boldsymbol{R} \times \frac{d\boldsymbol{P}}{dt}.$$

The first term is zero. Using Eq. (1) we get

$$\frac{d\boldsymbol{L}_0(CM)}{dt} = \boldsymbol{R} \times \boldsymbol{F}^{\text{ext}}. \tag{6}$$

It follows from Eqs. (5) and (6) that

$$\frac{d\boldsymbol{L}_{CM}^{\prime}}{dt} = \boldsymbol{N}_{CM}^{\prime\,\text{ext}}, \tag{7}$$

which is identical in form to the corresponding law in the laboratory frame.

S12 *I am surprised. The CM of a system may be accelerating; the CM frame, in general, is not an inertial frame. The rotational law of motion should not look the same as in an inertial frame.*

† The decomposition of $\boldsymbol{N}$, however, is true for any two frames. It involves taking moments of forces about two different origins and makes no use of the special nature of the CM frame.

T12 That is an important point. It is true that in the non-inertial CM frame there will be a pseudo-force $-m_1 d^2\boldsymbol{R}/dt^2$ on each particle. The total torque due to the pseudo-forces about the CM, however, sums up to zero since the pseudo-forces are proportional to masses (see Box 7.1). This is much like the fact that the weight of a body does not produce a torque about the CM of the body.

Box 7.1

Consider $d\boldsymbol{L}'_{CM}/dt$

$$\frac{d\boldsymbol{L}'_{CM}}{dt} = \frac{d}{dt}\sum_{i=1}^{N} m_i \boldsymbol{r}_i{}' \times \boldsymbol{v}_i{}'.$$

The CM frame has a purely translational motion with acceleration $d^2\boldsymbol{R}/dt^2$. The equation of motion for the ith particle in this non-inertial frame is

$$\frac{d\boldsymbol{p}_i{}'}{dt} = \boldsymbol{F}_i - m_i\frac{d^2\boldsymbol{R}}{dt^2}.$$

Hence

$$\frac{d\boldsymbol{L}'_{CM}}{dt} = \sum_{i=1}^{N} \boldsymbol{r}_i{}' \times \left(\boldsymbol{F}_i - m_i\frac{d^2\boldsymbol{R}}{dt^2}\right).$$

The second term is proportional to $\sum m_i\boldsymbol{r}_i{}'$ which is zero by the definition of the CM. Also in the first term the torques due to internal forces sum up to zero, using the strong form of the III law. Hence

$$\frac{d\boldsymbol{L}'_{CM}}{dt} = \boldsymbol{N}'^{\text{ext}}_{CM}.$$

Cylinder rolling down a plane

S13 *Can you illustrate these points with some example?*

T13 Consider a cylinder rolling down an inclined plane without slipping (see Fig. 7.1). The motion of the cylinder is clearly a combination of translation and rotation in the laboratory frame. In the CM frame the motion is purely rotational.

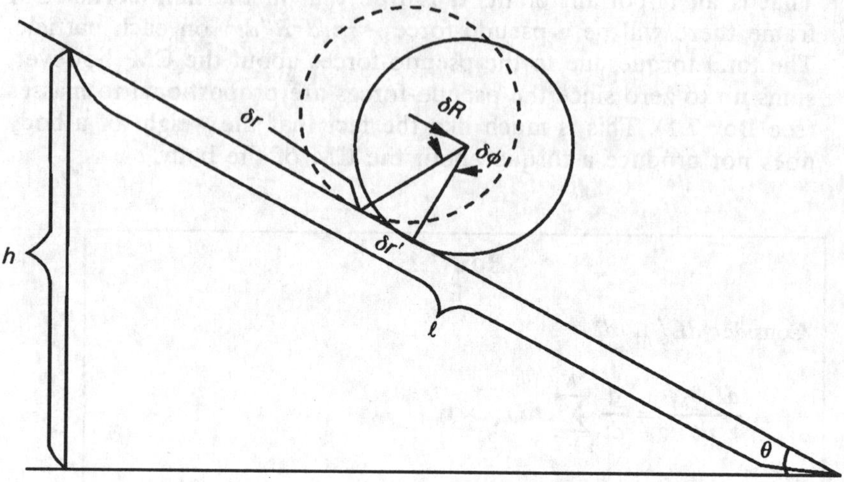

Fig. 7.1 Two instantaneous positions, a time interval δt apart, of a cylinder rolling down an inclined plane without slipping.

S14 *How do you put in the condition of rolling without slipping?*

T14 Rolling without slipping means there is no relative motion between the plane and the cylinder at the point of contact, i.e. in the laboratory frame the point on the cylinder in contact with the plane must have zero instantaneous velocity. Equation (2), therefore, implies that the displacement of the CM in the laboratory frame (δR) and the displacement of the point of contact in the CM frame ($\delta r'$) must add up to zero: $\delta R + \delta r' = 0$.

Since the cylinder has purely rotational motion in the CM frame, $|\delta r'| = c|\delta\theta|$ for a point on the rim of the cylinder, c being the radius of the cylinder. Therefore $|\delta R| = c|\delta\theta|$. Thus the magnitudes of the translational velocity and acceleration of the CM are related to the corresponding rotational quantities by $V = c\omega$ and $a_{CM} = c\alpha$, where α is angular acceleration.

S15 *This is the kinematics of the situation. How do you handle its dynamics?*

T15 Let us identify all the forces on the cylinder (see Fig. 7.2): The weight of the cylinder mg acts at its CM in the vertically downward direction. The normal force N by the inclined plane on the cylinder acts at the point of contact.

Since the cylinder rolls without slipping, there must be a force of friction f acting at the point of contact up the plane.

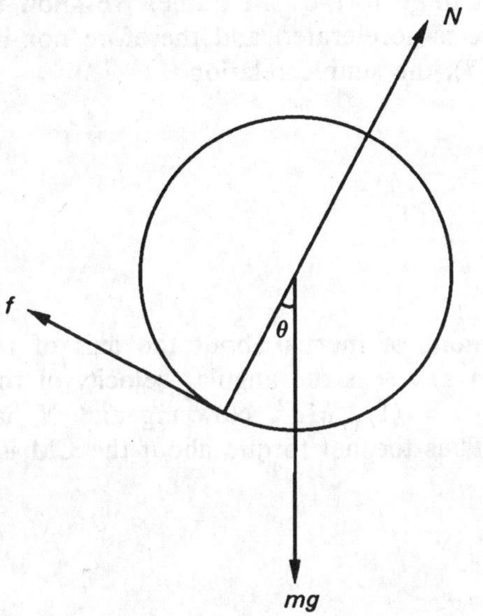

Fig. 7.2 Free-body diagram of the rolling cylinder.

S16 *Why do you say friction is necessary for rolling down a plane?*

T16 If there were no friction there would be no torque about the CM and therefore no rotational motion of the cylinder. The net force on the cylinder would be mg $\sin\theta$ acting at the CM which would produce a purely translational acceleration of the cylinder. The point of contact would therefore slide down with the same acceleration as that of the CM. To prevent this impending motion, friction acts in a direction up the plane and thereby provides a torque about the CM in the right sense, causing the rotation of the cylinder.

S17 *I get the point. Let us return to the calculation.*

T17 Since there is no motion perpendicular to the plane, the forces along that direction, mg $\cos\theta$ and N, must cancel each other. The net force acting down the inclined plane is mg $\sin\theta - f$. Therefore

$$mg\sin\theta - f = ma_{CM}, \tag{8}$$

where a_{CM} is the acceleration of the CM in the laboratory frame.

S18 *What about the rotational motion of the cylinder?*

T18 For this it is convenient to go to the CM frame. We know that, in our problem, this frame is accelerated and therefore non-inertial. Yet, as we saw in Eq. (7), the simple relation

$$\frac{dL'_{CM}}{dt} = N'^{\text{ext}}_{CM}$$

is true. Here

$$L'_{CM} = I_{CM}\omega,$$

where I_{CM} is the moment of inertia about the axis of rotation passing through the CM and ω is the angular velocity of rotation. For a solid cylinder $I_{CM} = (1/2)Mc^2$. Now mg and N have no torque about the CM. Thus the net torque about the CM is

$$N'_{CM} = fc.$$

We thus get

$$I_{CM}\alpha = fc, \tag{9}$$

where $\alpha = d\omega/dt$ is the angular acceleration of the cylinder. Equations (8) and (9) can be solved using the rolling condition $a_{CM} = c\,\alpha$ to obtain

$$a_{CM} = \frac{2}{3} g \sin\theta.$$

Evidently a_{CM} is less than $g\sin\theta$, the acceleration of a body sliding down without friction. (There is, however, no dissipation of energy due to friction since there is no relative motion between the two surfaces at the point of contact. See T30.)

S19 *You used the CM frame to obtain Eq. (9) for rotational motion. Can I solve the problem using the laboratory frame?*

T19 Of course you can. Let us take moments about the bottom of the inclined plane. The line of action of f passes through this point and hence produces no torque.† N and $mg\cos\theta$ being equal and opposite, and acting along the same line of action, produce no net torque either. Thus the only force that produces any torque about O is $mg\sin\theta$. Hence,

$$N_O = cmg\sin\theta.$$

† We are treating this as a two dimensional problem. The argument is valid for every slice of the cylinder.

Next,

$$L_0 = L'_{CM} + L_0(CM)$$
$$= I_{CM}\omega + cMV.$$

Hence,

$$\frac{dL_0}{dt} = I_{CM}\alpha + Mca_{CM}.$$

Using the rotational law of motion we get

$$cmg\sin\theta = I_{CM}\alpha + cMa_{CM}$$

which is the same equation as that obtained from Eqs. (8) and (9).

Work–Energy Theorem

S20 *Let us move on to work–energy theorem. How does it get generalized to a system of N particles?*

T20 This is an important topic and one must appreciate a number of points regarding it.

Remember that each particle in the system is subject to a force which is a sum of an external force and an internal force due to the other particles in the system. As the system moves from one configuration to another, the total work done on the system is the sum of the work done by the force on each particle.

$$\Delta W = \sum_{i=1}^{N} \int_1^2 \boldsymbol{F}_i \cdot d\boldsymbol{r}_i \tag{10}$$

where

$$\boldsymbol{F}_i = \boldsymbol{F}_i^{ext} + \sum_{\substack{j=1 \\ j \neq i}}^{N} \boldsymbol{F}_{ij}$$

S21 *Doesn't the work done by the internal forces add up to zero because of the III law?*

T21 This is a possible pitfall. The internal forces do add up to zero because of the III law, which is why we get the simple equation $d\boldsymbol{P}/dt = \boldsymbol{F}^{\text{ext}}$ for a many particle system. This, however, does not mean that the work done by the internal forces also adds up to zero. Beware against this pseudo-III law. We write Eq. (10) as

$$\Delta W = \sum_{i=1}^{N} \int_1^2 \boldsymbol{F}_i^{\text{ext}} \cdot d\boldsymbol{r}_i + \sum_{\substack{i\,j=1 \\ i \neq j}}^{N} \int_1^2 \boldsymbol{F}_{ij} \cdot d\boldsymbol{r}_i$$
$$= \Delta W^{\,ext} + \Delta W^{\,int}.$$

The displacements in the above equation may in general be different for different particles. 'Therefore $\Delta W^{\,int}$ is non-zero even though $\boldsymbol{F}_{ij} = -\boldsymbol{F}_{ji}$. †

Now,

$$\boldsymbol{F}_i = m_i \frac{d\boldsymbol{v}_i}{dt}.$$

Therefore,

$$\sum_{i=1}^{N} \int \boldsymbol{F}_i \cdot d\boldsymbol{r}_i = \sum_{i=1}^{N} \int \frac{d}{dt}\left(\frac{1}{2} m_i \boldsymbol{v}_i^2\right) dt = \sum_{i=1}^{N} \Delta T_i,$$

where ΔT_i is the kinetic energy of the ith particle. Thus

$$\Delta T = \Delta W^{\text{ext}} + \Delta W^{\,int}, \tag{11}$$

where T is the total kinetic energy of the system. This is the work–energy theorem for a system of particles.

S22 *What does the work–energy theorem look like in the CM frame?*

T22 In order to see its form in the CM frame, let us decompose ΔW^{ext} and $\Delta W^{\,int}$ like we did other quantities in T8 and T10.

From Eq. (2), we have

$$\delta \boldsymbol{r} = \delta \boldsymbol{R} + \delta \boldsymbol{r}'.$$

---

† For a rigid body, however, the work done by the internal forces is always zero. This is because the general motion of a rigid body is a combination of rotation and translation. For rotation $d\boldsymbol{r}_i$ is normal to $\boldsymbol{F}_i^{\text{int}}$; hence no work is done. For translation, $d\boldsymbol{r}_i$ is common to all particles, so ΔW^{int} is zero because of the III law.

Therefore

$$\begin{aligned}\Delta W^{\text{ext}} &= \sum_{i=1}^{N}\int_1^2 \boldsymbol{F}_i^{ext}\cdot d\boldsymbol{r}_i \\ &= \int_1^2\left(\sum_{i=1}^{N}\boldsymbol{F}_i^{ext}\right)\cdot d\boldsymbol{R} + \sum_{i=1}^{N}\int_1^2 \boldsymbol{F}_i^{ext}\cdot d\boldsymbol{r}_i' \\ &= \int_1^2 \boldsymbol{F}^{ext}\cdot d\boldsymbol{R} + \Delta W'^{\,ext}.\end{aligned} \tag{12}$$

The first term in the equation above denotes the work done by the total external force on the CM. The second term is the total work done by the external forces in the CM frame. The first term is equal to the change in kinetic energy of the CM. As you can expect, the proof is identical to that in the single particle case:

We have $\quad \boldsymbol{F}^{\,ext} = M\dfrac{d\boldsymbol{V}}{dt},$

where M is the total mass of the system and $\boldsymbol{V}$ is the velocity of the CM. Therefore

$$\int_1^2 \boldsymbol{F}^{\text{ext}}\cdot d\boldsymbol{R} = \int_1^2 M\frac{d\boldsymbol{V}}{dt}\cdot \boldsymbol{V}dt = \int_1^2 d\left(\frac{1}{2}MV^2\right) = \Delta T(\text{CM}). \tag{13}$$

A similar decomposition of ΔW^{int} yields

$$\Delta W^{\text{ int}} = \int_1^2\left(\sum_{i=1}^{N}\boldsymbol{F}_i^{int}\right)\cdot d\boldsymbol{R} + \sum_{i=1}^{N}\int_1^2 \boldsymbol{F}_i^{\text{int}}\cdot d\boldsymbol{r}_i'$$

However, here the first term is zero because of the III law. The second term is the work done by the internal forces in the CM frame. Hence,

$$\Delta W^{int} = \Delta W'^{\text{ int}}. \tag{14}$$

We can appreciate this equation in another way:

$$\begin{aligned}\Delta W^{\text{ int}} &= \sum_{\substack{ij=1\\ i\neq j}}^{N}\int \boldsymbol{F}_{ij}\cdot d\boldsymbol{r}_i = \frac{1}{2}\sum_{\substack{ij=1\\ i\neq j}}^{N}\int \boldsymbol{F}_{ij}\cdot d\boldsymbol{r}_{ij} \\ &= \frac{1}{2}\sum_{\substack{ij=1\\ i\neq j}}^{N}\int \boldsymbol{F}_{ij}\cdot d\boldsymbol{r}_{ij}' = \Delta W'^{\text{ int}},\end{aligned}$$

where we have used the fact that the change in the relative vector is the same in the two frames; $\delta \boldsymbol{r}_{ij} = \delta \boldsymbol{r}'_{ij}$.

We use Eqs. (12), (14) and (4) in the work-energy theorem, Eq. (11), to obtain

$$\int_1^2 \boldsymbol{F}^{\text{ext}} \cdot d\boldsymbol{R} + \Delta W'^{\text{ext}} + \Delta W'^{\text{int}} = \Delta T(\text{CM}) + \Delta T',$$

which gives, because of Eq. (13),

$$\Delta T' = \Delta W' = \Delta W'^{ext} + \Delta W'^{\text{int}}. \tag{15}$$

This looks identical in form to Eq. (11) in the laboratory frame.

S23 *That surprises me. The CM frame is, in general, a non-inertial frame. Why do we get the same form of the work–energy theorem as in an inertial frame?*

T23 This is a non-trivial point and the answer lies both in the way the CM frame is defined and the fact that the pseudo-forces of a non-inertial frame are proportional to the masses. The net result is that the total work done by the pseudo-forces in the CM frame is zero.

The equation of motion for the ith particle in the CM frame is

$$\frac{d\boldsymbol{p}'_i}{dt} = \boldsymbol{F}_i - m_i \frac{d^2\boldsymbol{R}}{dt^2},$$

where the second term involving the acceleration $d^2\boldsymbol{R}/dt^2$ of the CM is the pseudo-force on the ith particle. The total work done in the CM frame is, therefore, given by

$$\Delta W' = \sum_{i=1}^{N} \int_1^2 \boldsymbol{F}_i \cdot d\boldsymbol{r}'_i = \sum_{i=1}^{N} \int_1^2 \left(\frac{d\boldsymbol{p}'_i}{dt} + m_i \frac{d^2\boldsymbol{R}}{dt^2} \right) \cdot d\boldsymbol{r}'_i. \tag{16}$$

The last term contains $\sum m_i d\boldsymbol{r}'_i$, which is zero since $\sum m_i \boldsymbol{r}'_i = 0$ by the definition of the CM. Also,

$$\sum_{i=1}^{N} \int_1^2 \frac{d\boldsymbol{p}'_i}{dt} \cdot d\boldsymbol{r}'_i = \sum_{i=1}^{N} \int_1^2 m_i d\boldsymbol{v}'_i \cdot \boldsymbol{v}'_i$$

$$= \int_1^2 d\left(\sum_{i=1}^{N} \frac{1}{2} m_i v_i'^2 \right) = \Delta T' \tag{17}$$

From Eqs. (16) and (17), we get the work–energy theorem in the CM frame as stated in Eq. (15).

Note that, in the CM frame, the displacements of all particles are not independent: $\sum m_i d\boldsymbol{r}_i' = 0$.

Do the roles of external and internal forces separate?

S24 *Am I right in thinking that the work done by the external forces changes the kinetic energy of the CM and that by the internal forces changes the kinetic energy of relative motion, i.e. the kinetic energy in the CM frame?*

T24 No, that is not generally so. Do not confuse 'work done by the sum of forces' with 'the sum of work done by individual forces'. The two quantities are different because the displacements of different particles may be different. Thus the work done by the total external force is equal to the change in the kinetic energy of the CM. However, this is not the same as the total work done by external forces which contributes in general to both the change in CM energy and the change in energy of relative motion. This is evident from Eqs. (12) and (13). This also shows that the second part of your statement is not correct. The change in kinetic energy of relative motion arises both from $\Delta W'^{\,\text{int}}$ and $\Delta W'^{\,\text{ext}}$. In other words the simple-minded separation that you suggest is not correct.

Such a separation, however, does occur if $\Delta W'^{\,\text{ext}} = 0$. In this case $\Delta T(CM) = \int \boldsymbol{F}^{\text{ext}} \cdot d\boldsymbol{R}$ and $\Delta T' = \Delta W^{\text{int}}$. You will encounter examples of this kind later.

S25 *Suppose the total external force on a system is zero. In this case is it correct to say that the change in kinetic energy of the system is due only to the work done by the internal forces?*

T25 No. This is a similar error to the one that we discussed above. If the total external force is zero, it does not mean total work done by the external forces is zero. $\boldsymbol{F}^{\text{ext}} = 0$ implies that the velocity or the kinetic energy of the CM does not change. But $\Delta W'^{\,\text{ext}}$ may well be non-zero. Thus even when $\boldsymbol{F}^{\text{ext}}$ is zero, the kinetic energy of the system may change due to both the work done by the internal and that by the external forces.

S26 *What if the total work done by the external forces is zero?*

T26 This is a very interesting case. From Eq. (11), in this case,

$$\Delta T = \Delta T(CM) + \Delta T' = \Delta W^{\text{int}}.$$

So the work done by the internal forces equals the change in kinetic energy of the CM plus the change in kinetic energy of relative motion. Note, however, that the kinetic energy of the CM can change only due to a net external force. We have here a situation opposite to that in *S*25; ΔW^{ext} is zero but $\boldsymbol{F}^{\text{ext}}$ may not be zero, so that by Eq. (12)

$$\Delta W'^{\text{ext}} = -\int \boldsymbol{F}^{\text{ext}} \cdot d\boldsymbol{R} = -\Delta T(CM).$$

This curious situation is in fact encountered frequently as will be seen later.

S27 *Can you put together the various points regarding the work–energy theorem discussed so far?*

T27 1. The change in kinetic energy of a system arises due to both the work done by the external forces and the work done by the internal forces. [Eq. (11)]

2. The total work done by a number of forces is not the same as the work done by the total force in moving the CM, since individual displacements may be unequal. This is why the work done by internal forces is not necessarily zero even though the internal forces always add up to zero. Likewise, total work done by the external forces is not in general equal to the work done by the total external force. (The latter is equal to the change in the kinetic energy of the CM.) There is an additional term representing the work done by the external forces in the CM frame [Eq.(12)].

3. It follows that, if the total external force is zero, the kinetic energy of the CM (T(CM)) does not change. The kinetic energy of the system (T) may, however, change; it appears as a change in the kinetic energy of relative motion (T'). Furthermore, the work done by both the internal as well as the external forces can contribute to this change.

4. Interestingly, the kinetic energy of the CM can change even when ΔW^{ext}—the total work done by the external forces is zero. Obviously the total external force itself must not be zero for this to happen. (It so happens in this case that $\int \boldsymbol{F}^{\text{ext}} \cdot d\boldsymbol{R}$—the work done by the total external force on the CM of the system and $\Delta W'^{\text{ext}}$—the work done by the external forces in the CM frame have equal magnitude and opposite signs, so they add up to zero.)

5. The work–energy theorem has the same form in the CM frame as in the laboratory frame even though the CM frame may be a non-inertial frame. This happens because the work done by the pseudo-forces in the CM frame adds up to zero.

S28 *Can you illustrate these points with some examples?*

T28 Consider a rigid body falling freely under gravity. Here the displacements of all the particles of the body are equal, i.e. the displacements are zero in the CM frame. In this case

$$\Delta W^{\text{int}} = \Delta W'^{\,\text{int}} = 0,$$

$$\Delta W'^{\,\text{ext}} = 0, \ \Delta T' = 0,$$

$$\Delta T = \Delta T(\text{CM}) = \Delta W^{\text{ext}}.$$

For a rigid body that has translational as well as rotational motion (e.g., a cylinder rolling down an inclined plane) the displacements in the CM frame are not zero. $\Delta W'^{\,\text{int}}$, however, continues to be zero because the displacements in the CM frame are all perpendicular to the internal forces. In this case

$$\Delta W^{\text{int}} = \Delta W'^{\,\text{int}} = 0,$$

$$\Delta T(\text{CM}) = \int \boldsymbol{F}^{\text{ext}} \cdot d\boldsymbol{R},$$

$$\Delta T' = \Delta W'^{\,\text{ext}}.$$

Let us look at an example wherein the kinetic energy of a system changes due to internal forces alone. Consider two masses connected by a compressed spring. When the spring is released, the work done by the internal forces due to the spring increases the kinetic energy of the system. Notice that the internal forces add up to zero while the work done by them does not. Since there is no external force on the system, the kinetic energy of the CM of the system does not change. In this case

$$\Delta W^{\text{ext}} = 0, \quad \Delta W'^{\,\text{ext}} = 0,$$

$$\Delta T(\text{CM}) = 0,$$

$$\Delta T = \Delta T' = \Delta W^{\,\text{int}}.$$

Next suppose that the above mass-spring system is allowed to fall under gravity as the spring is released. In this case

$$\Delta W'^{\,\text{ext}} = \int m_1 g\hat{\boldsymbol{k}} \cdot d\boldsymbol{r}_1' + \int m_2 g\hat{\boldsymbol{k}} \cdot d\boldsymbol{r}_2'$$

$$= g\hat{\boldsymbol{k}} \cdot \int d(m_1\boldsymbol{r}_1' + m_2\boldsymbol{r}_2') = 0,$$

where $\hat{\boldsymbol{k}}$ is the direction of gravity. Thus

$$\Delta W^{\text{ext}} = \int \boldsymbol{F}^{\text{ext}} \cdot d\boldsymbol{R} = \Delta T(CM)$$

$$\Delta T' = \Delta W^{\text{int}} = \Delta W'^{\,\text{int}}$$

There is a clear separation here. The work done by the external gravitational forces changes the kinetic energy of the CM, while the work done by the internal spring forces changes the kinetic energy of relative motion.

S29 *Are there examples where this kind of separation does not hold?*

T29 As remarked in T24, this separation will not occur whenever $\Delta W'^{\,\text{ext}}$ is not zero. A simple example is when two masses connected by an ideal spring are subject to two equal and opposite external forces. At each mass, the external force is balanced by the internal spring force. In this case

$$\Delta W^{\text{ext}} = \Delta W'^{\,\text{ext}} = -\Delta W'^{\,\text{int}},$$

$$\Delta T(CM) = 0, \qquad \Delta T' = 0.$$

Since $-\Delta W'^{\,\text{int}}$ equals change in the potential energy of the spring, the work of the external forces in this example changes the internal potential energy of the system.

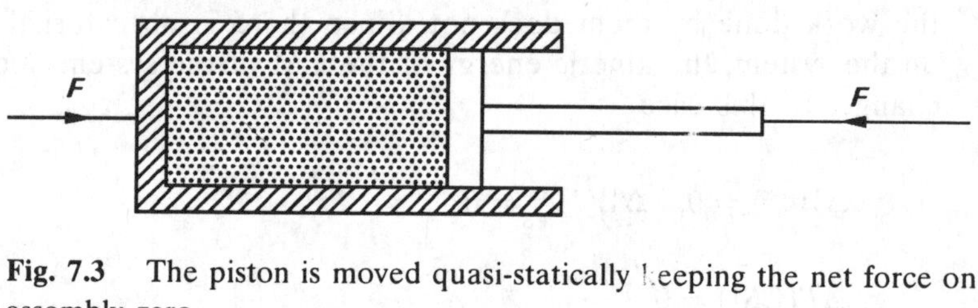

Fig. 7.3 The piston is moved quasi-statically keeping the net force on the assembly zero.

As another example, consider a gas enclosed in a cylinder with a movable piston. If we compress the gas quasi-statically, i.e. always keeping it in equilibrium, the total external force on the system (gas) is zero† (See Fig. 7.3). The total work done by the external forces, however, is not zero since the point of application of only one of the forces moves. This means that $\Delta W'^{\,\text{ext}}$ is not zero. For an ideal gas there are no internal forces, so $\Delta W'^{\,int}$ is zero. (If the gas is not ideal, $-\Delta W'^{\,int}$ equals the change in its internal potential energy.) Thus $\Delta W'^{\,\text{ext}}$ equals the change in the internal energy U of the gas (i.e., energy in its CM frame), which includes both its kinetic energy and potential energy:

$$\Delta W^{\text{ext}} = \Delta W'^{\,\text{ext}} = \Delta T' - \Delta W'^{\,\text{int}} = \Delta T' + \Delta V' = \Delta U.$$

Rolling cylinder revisited

S30 *Shall we look at some specific problem in greater detail?*

T30 Let us return to the problem of a cylinder rolling down an inclined plane. We have already seen the dynamics of the problem earlier (T13 to T19). Now let us look at it from the point of view of the work–energy theorem that we have been discussing about.

Let the distance travelled by the cylinder along the inclined plane be ℓ, so that the CM of the cylinder falls through a vertical height $h = \ell \sin\theta$. See Fig. 7.1. Since the cylinder is a rigid body, the internal forces in this example do not do any work, i.e. $\Delta W^{\text{int}} = \Delta W'^{\,\text{int}} = 0$.† There are three external forces: the weight, the normal force and the force of friction. We want to calculate $\Delta T(CM)$ and $\Delta T'$. The work done by the weight = mgh.

Since the cylinder rolls without slipping, there is no relative motion between the cylinder and the plane at the point of contact. (The force between the two surfaces is the force of static friction.) The point of application of the force does not undergo any displacement and, hence, the work done by f is zero.

The work done by the normal force is also zero since it also acts at the point of contact which is instantaneously stationary.

Thus the total work done by the external forces is

$$\Delta W^{\text{ext}} = mgh.$$

---

† We are applying the work–energy theorem to the gas as an isolated system. This means that the process is adiabatic.

† See footnote on page 198.

The net external force on the cylinder is $mg\sin\theta - f$. The work done by this force is $(mg\sin\theta - f)\ell$. Notice that, for calculating the work done by the net external force, we have used the displacement of the CM: i.e.

$$\int_1^2 \boldsymbol{F}^{\text{ext}} \cdot d\boldsymbol{R} = (mg\sin\theta - f)\ell.$$

Evidently the total work done by the external forces is not equal to the work done by the net external force. This is something we emphasized in T24 and T25. From Eq. (12) it follows that $\Delta W'^{\,\text{ext}} = f\ell$.

Using Eqs. (13) and (15) we get

$$\Delta T(CM) = (mg\sin\theta - f)\ell, \qquad \Delta T' = f\ell.$$

We are still not through; we have to find f. For this, let us use the kinematic definitions of $\Delta T'$ and $\Delta T(CM)$:

$$\Delta T(CM) = \frac{1}{2}M(v_f^2 - v_i^2),$$

$$\Delta T' = \frac{1}{2}I(\omega_f^2 - \omega_i^2),$$

so that Eq. (11) gives

$$mgh = \frac{1}{2}M(V_f^2 - V_i^2) + \frac{1}{2}I(\omega_f^2 - \omega_i^2). \tag{18}$$

Substituting in this equation the condition for rolling without slipping, $V = c\omega$, we get

$$\Delta T' = \frac{1}{2}I(\omega_f^2 - \omega_i^2) = \frac{mgh}{3}.$$

A comparison of this expression with $\Delta T' = f\ell$, obtained earlier gives

$$f = \frac{1}{3}mg\sin\theta.$$

Thus

$$\Delta T(CM) = \frac{2}{3}mgh; \quad \Delta T' = \frac{1}{3}mgh.$$

S31 *Are these results consistent with those obtained earlier by the dynamical method?*

T31 They must be; the work–energy theorem is a consequence of the dynamical equations. Let us verify.

We have obtained $a_{CM} = (2/3)g\sin\theta$. Obviously, therefore

$$\Delta T(CM) = \frac{1}{2}M(V_f^2 - V_i^2) = Ma_{CM}\ell = \frac{2}{3}Mg\ell\sin\theta.$$

Also, the angular acceleration $\alpha = a_{\text{CM}}/c$. Hence, the change in the kinetic energy in the CM frame can be calculated as

$$\Delta T' = \frac{1}{2}I(\omega_f^2 - \omega_i^2) = I\alpha\Phi = \frac{1}{3}Mg\sin\theta(c\phi),$$

where Φ is the angular displacement of the cylinder in rolling down the plane. Clearly, $\ell = c\Phi$ and, hence,

$$\Delta T' = \frac{1}{3}Mg\ell\sin\theta.$$

These are the same results that we obtained from the work–energy theorem.

Notice that only a part of ΔW^{ext} goes into increasing the kinetic energy of the CM. Part of it is responsible for the increase in the kinetic energy in the CM frame (i.e. rotational energy).

S32 *I notice that Eq. (18) is just the principle of conservation of energy (kinetic energy + potential energy) applied to the problem of a rolling cylinder. Is it correct to use it even when there are non-conservative forces such as friction and normal reaction in the problem?*

T32 In general not. See Eq. (7) Ch.6. In this particular problem, however, the work done by f and N is zero since their point of application is stationary. This is why the conservation principle still gives the right answer.

S33 *I am puzzled by one thing. You got in T30 that $\Delta W'^{\text{ext}} = f\ell$. This looks like the work done by the force of friction. But you just said that the force of friction does not do any work in this example.*

T33 You must realize that work is not invariant under transformation from one frame to another in relative motion. The work done by the frictional force in the laboratory frame is zero; however, it is not zero in the CM frame. In the CM frame the instantaneous point of contact has non-zero velocity (due to the rotation of the cylinder). The plane below also has the same velocity so that the relative velocity between them is zero. The point of application of the force of friction thus undergoes infinitesimal displacement which is the same as the infinitesimal displacement of the plane below. Over the entire motion, therefore, the work done by the frictional force is $f\ell$.

The work done by the normal force continues to be zero in this frame also since the displacement of the point of contact is perpendicular to the force. The CM, which is the point of application of the force of gravity, is, by definition, at rest in the CM frame. Obviously, therefore, the weight does no work in the CM frame. All put together, the total work done in the CM frame by the external forces, $\Delta W'^{ext} = f\ell$.

Often $\Delta W_f'$ is expressed slightly differently:

$$\Delta W_f' = \int_1^2 f dr' = \int_1^2 f c d\phi.$$

Thus, the frictional work is also viewed as the work done by the frictional torque fc.

The role of internal forces in the overall motion of a system

S34 *In these examples work done by external forces contributes to change in internal energy, i.e. energy in the CM frame. Can we have the other way around: the work done by internal forces changing the kinetic energy of the CM?*

T34 It may surprise you, but situations of this type are quite commonplace. Indeed this is precisely the situation we have described in T26 and (4) of T27. We must caution again that in order to have a change in the kinetic energy of the CM of the system we must have a non-zero $\boldsymbol{F}^{\text{ext}}$; but ΔW^{ext} may be zero or even negative.

A nice example of this situation is the mass-spring assembly that we considered in Ch.4. Consider the same system with one difference; imagine that the upper plate of the system (see Fig. 7.4) is displaced through a distance X and constrained from moving from that position (either by tying a string or by inserting a rod if X is negative). Let us determine the minimum $|X|$ for which the assembly jumps off the ground when the constraining string (or rod) is removed suddenly.

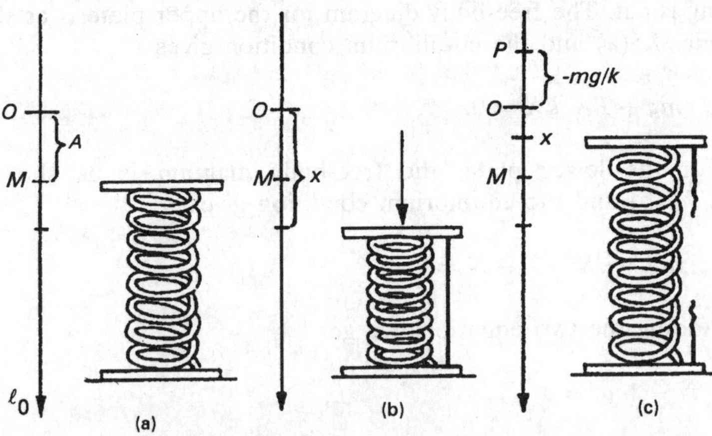

Fig. 7.4 Self-hopping mass-spring assembly. (a) The spring is compressed under the weight of upper plate. (b) It is compressed further and tied by a string. (c) An instantaneous configuration of the assembly after the string is cut.

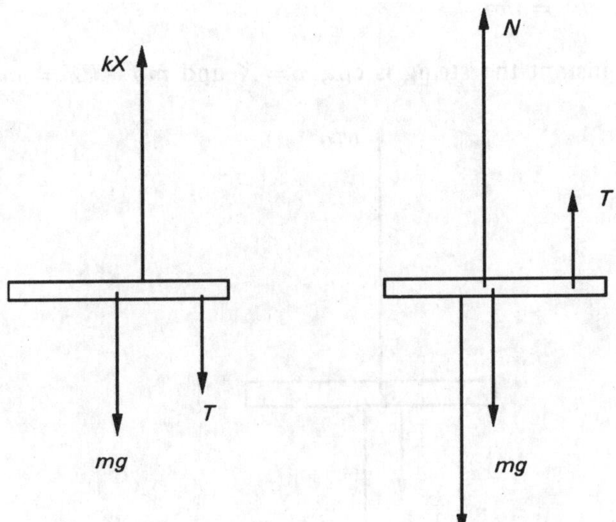

Fig. 7.5 The mass-spring system: situation in Fig. 7.5(b). Free-body diagram (a) for the upper plate and (b) for the lower plate.

To begin with let us consider the equilibrium situation before the string is cut. The free-body diagram for the upper plate is as shown in Fig. 7.5(a) and the equilibrium condition gives

$$mg + T - kX = 0.$$

For the lower plate, the free-body diagram is as shown in Fig. 7.5(b) and the equilibrium condition gives

$$mg + kX - T - N = 0.$$

Adding the two equations we get

$$N - 2mg = 0$$

which, as expected, describes the equilibrium condition for the entire system.

When the constraining force is removed suddenly (i.e. if the string is cut), the upper plate is no more in equilibrium. If x denotes the instantaneous position of the upper plate, the equation of motion for the upper plate may be written as

$$mg - kx = ma. \tag{19}$$

At the instant the string is cut, $x = X$ and $mg - kX = ma$.

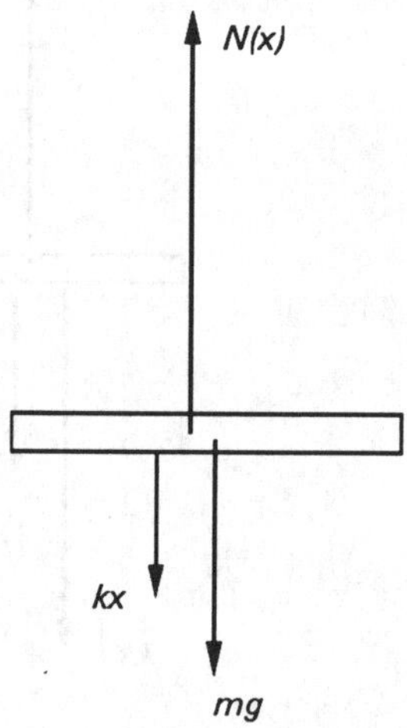

Fig. 7.6 Free-body diagram for the lower plate after the string is cut.

Because of the presence of the self-adjusting force of normal reaction, the lower plate continues to be in equilibrium as long as

N is non-negative. The free-body diagram for the lower plate now is as given in the Fig. 7.6 and the II law gives

$$N(x) = mg + kx.$$

Thus, as long as $x \geq -mg/k \equiv -A$, $N \geq 0$ and the lower plate remains in equilibrium. (Note that we have taken the downward direction to be positive throughout.)

S35 *But how do you determine the minimum $|X|$ for which the assembly jumps?*

T35 The equilibrium position of the upper plate under the action of its weight and the spring force alone will be $x = A = mg/k$. So, when compressed as above and released, the upper plate oscillates about this mean position with an amplitude $X - A$. As long as this amplitude is small enough, i.e. as long as $X - A \leq 2A$, $x \geq -A$ and the assembly will not jump. The minimum compression necessary for the assembly to jump is, thus, $X = 3A$. The minimum extra compression that is needed is thus $2A$. This is the same result that we obtained in Ch.2.

S36 *What does all this have to do with the work–energy theorem?*

T36 So far nothing. Let us now verify the work–energy theorem in this example. First note that the system accelerates in the upward direction only during the interval from the instant the string is cut to the instant it takes off the ground. Once off the ground, the motion of the CM of the assembly is governed by the only external force acting on the system—the force of gravity.

Let us obtain $x(t)$ for the interval during which the system accelerates in the upward direction. Equation (19) can be written as

$$m\frac{d^2x}{dt^2} + kx = mg.$$

Its solution has the form

$$x(t) = A + C\cos(\omega t + \alpha).$$

where $\omega^2 = k/m$ and $A = mg/k$. The initial conditions $x = X$ and $v = dx/dt = 0$ at $t = 0$ imply $\alpha = 0$ and $C = (X - A)$. Hence,

$$x(t) = A + (X - A)\cos\omega t.$$

so that the velocity of the upper plate is

$$v(t) = -(X - A)\omega \sin \omega t.$$

Let the instant the system jumps off the ground be denoted by t_f. Since the normal force N is zero at this instant, we have

$$\begin{aligned} N(t_f) &= mg + kx(t_f) \\ &= 2mg + k(X - A)\cos \omega t_f = 0 \end{aligned}$$

Now, the lower plate is at rest throughout the time interval from $t = 0$ to $t = t_f$. The change in the kinetic energy of the system, therefore, is the same as the change in the kinetic energy of the upper plate, which in turn is just its final kinetic energy. Hence,

$$\Delta T = \frac{1}{2}k(X - A)^2 \sin^2 \omega t_f.$$

Let us calculate the work done by the forces. Since the lower plate does not move throughout, ΔW^{ext} is just the work done by the weight of the upper plate:

$$\begin{aligned} \Delta W^{\text{ext}} &= \int_i^f mg dx \\ &= -mg(X - A)(1 - \cos \omega t_f) \\ &= -mg(X + A). \end{aligned}$$

The work done by the internal forces is again just the work done by the spring force on the upper plate:

$$\begin{aligned} \Delta W^{\text{int}} &= -\int_i^f kx dx \\ &= \frac{1}{2}k(X - A)^2 \sin^2 \omega t_f + mg(X + A). \end{aligned}$$

Thus, we see that $\Delta W^{\text{ext}} + \Delta W^{\text{int}} = \Delta T$, which is the work–energy theorem.

Notice that $\Delta W^{\text{ext}} < 0$. Thus the positive work done by the internal forces not only provides for the increase in the kinetic energy of the system, it also compensates for the negative work done by the external forces.

S37 *Now I see what you mean by internal forces providing for the energy of the overall motion of a system. But what if I consider the system as equivalent to a point mass 2m located at the CM?*

T37 In that case you will describe only the CM motion of the system, which is determined entirely by the net external force on the system. In the example above we can readily see that the work done by the net external force on the CM equals the increase in its kinetic energy.

The CM of the system is located at $x_{CM} = (x + \ell_0)/2$ and the total mass of the system is 2m. So the change in kinetic energy of the CM is

$$\Delta T(CM) = \frac{1}{2}(2m)\left(\frac{dx_{CM}}{dt}\right)^2 = \frac{1}{4}k(X - A)^2 \sin^2 \omega t_f.$$

The net force on the CM is

$$\begin{aligned} F^{ext} &= 2mg - N \\ &= -k(X - A)\cos\omega t. \end{aligned}$$

Therefore, the work done by the net external force on the CM is

$$\begin{aligned} \int_i^f F^{ext} dx_{CM} &= \int_0^{t_f} [-k(X - A)\cos\omega t][-\frac{1}{2}\omega(X - A)\sin\omega t dt] \\ &= \frac{1}{4}k(X - A)^2 \sin^2 \omega t_f \\ &= \Delta T(CM) \end{aligned}$$

which verifies the assertion.

In elementary problems, imagining the system mass to be concentrated at its CM, i.e., looking at the CM motion alone, is often adequate. However, in many situations, especially those involving non-rigid systems, this picture may cloud the real physics of the problem.

For example, we often explain the motion of a car, or of a person walking or running, as caused by the external force (static friction) exerted by the ground on the object. To the extent that we consider the motion of the CM of the objects, it is the external force which causes the motion. It is entirely wrong, however, to think that the work done by this external force brings about the change in the kinetic energy of the objects. In the examples cited, the work done by the external force is zero since its point of application does not move. (If we take into account the fact that the contact between car tyres and road, or that between feet and ground, is never really perfect and there always are some frictional losses, the external work, in fact, is negative.) It is the internal forces which have to perform positive work to provide for the increase in the kinetic energy of the objects. This is the reason why fuel needs to be burnt <u>inside</u> a car, or a man gets tired.

S38 *This is an interesting point. Any other example?*

T38 Yes, consider a man climbing up a pole. How does he climb? He holds on to the pole with both his hands and pulls his body up through a certain distance. Then he anchors his body in this position with his feet, repositions his hands at a higher position and once again pulls his body up. Repeating this action over and over again he gets up the pole. His motion, thus, involves repetitive cycles of acceleration, deceleration and momentary rest. The external forces on the man are: his weight acting in the downward direction and the frictional force exerted by the pole acting in the upward direction. Let us understand this motion in terms of the work–energy theorem.

Consider the forces acting on the hands when the body is being pulled up. (The hands are being considered as the 'free-body'.) The hands are stationary at the instant the man is rising, so the net force on the hands must be zero. The force exerted by the pole acts on the hands in the upward direction (to oppose the weight of the man and keep him from sliding down). Clearly the rest of the body must exert a force on the hands in the downward direction such that the net force on the hands is zero. Now, in accordance with Newton's III law, the hands exert an equal and opposite force on the rest of the body. This force, acting in the upward direction, is responsible for the upward motion of the body.

Notice that the point of application of the force exerted by the pole does not move; the work done by this force is zero. The work done by the weight is negative since the CM of the man moves upward. Therefore the total work done by the external forces is negative: $\Delta W^{\text{ext}} < 0$. If the man moves up with average constant velocity, ΔT is zero. If he moves with a non-zero average acceleration, ΔT is positive. In either case, from the work–energy theorem, $\Delta W^{\text{int}} > 0$. The total work done by the internal forces, thus is positive and is the source of energy of motion of the man. The man needs to expend internal energy to generate this work, which is why climbing up a pole tires him.

8

Rigid Bodies

Rotation and Translation

S1 *What does one understand by a 'rigid body'?*

T1 A rigid body is a system of particles in which the distance between any two particles is constrained to be fixed; i.e. r_{ij}= constant for every pair of particles indicated by $\{i, j\}$. This means, for any arbitrary motion of a rigid body, the relative displacement between any two particles $\delta \boldsymbol{r}_{ij}$ must be normal to the relative vector $\boldsymbol{r}_{ij}$. The internal forces, therefore, do no work; the internal potential energy of a rigid body remains constant and, therefore, can be ignored in the study of rigid body motion. All the general results of the previous chapter are applicable to this case. However, it is possible, and proves to be a great simplification, to specialize those results for rigid bodies, using the constraints mentioned above.

S2 *What is the simplification achieved?*

T2 The rigid body constraints imply that the number of co-ordinates required to specify the locations of all the particles of a rigid body is only six. This is intuitively obvious. We can choose a point in the rigid body as the origin and fix three independent axes in the body. Three co-ordinates are required to specify the location of the origin and three angles are required to specify the orientation of the body-fixed axes relative to the space-fixed axes (i.e. the axes of the laboratory frame).† With the location of the origin and the orientation of the body-fixed axes specified, these constraints determine the locations of all the other particles of the rigid body. This is an enormous simplification, since for a general system of N particles, the number of degrees of freedom is $3N$. (For a physical body of ordinary size N is of the order of Avogadro's number.)

---

† A convenient choice of the three angles of rotation needed to go from the laboratory frame to the body-fixed frame are the so-called Euler's angles. The reader may look up any standard text for a definition of these angles.

S3 *How do we view the general motion of a rigid body?*

T3 The displacement of a rigid body with one point fixed is always a rotation about some axis. (The axis may not be fixed; its orientation may be continuously changing with time.) This intuitively obvious statement can be proved rigorosly and is known as Euler's theorem.

This means that the most general displacement of a rigid body can be viewed as a translation of an arbitrary point associated with the body†† plus a rotation of the body about that point.

Note one significant point. In the previous chapter, we saw that for a many particle system Newton's laws yield two equations:

$$\frac{d\boldsymbol{P}}{dt} = \boldsymbol{F}^{\text{ext}}; \ \frac{dL}{dt} = \boldsymbol{N}^{\text{ext}}. \tag{1}$$

For a general N particle system (with $N > 2$) these equations provide only a partial description of motion. For a rigid body with six degrees of freedom, however, the solution of these two vector equations constitutes a complete solution of the problem.

S4 *When does one say that a rigid body has pure rotation?*

T4 A rigid body has purely rotational motion if at least one of the points associated with the body is fixed in space (i.e. fixed in the laboratory frame of reference). Rotation can be about a fixed axis or about a fixed point.

A rotation about a fixed axis, for example a flywheel rotating about its axis, is simple to analyze. All the points on the axis of rotation are stationary in the laboratory frame (see Fig. 8.1a).

A rotation about a fixed point, for example the motion of a spinning top, by contrast, is far more complex. You must have observed how a top usually moves about on the floor, as it rotates, before settling down in a steady motion at one point. Then on, its point of contact with the ground remains fixed and the ensuing motion of the top is purely rotational. It spins about its symmetry axis. The axis, however, is not fixed; it keeps moving at an angle, as shown in Fig. 8.1(b), around the vertical passing through the fixed point of contact. This is called the precession of the top. Moreover, the inclination of the axis with the vertical goes on changing with time—the related motion is called *nutation.*

†† This point need not 'belong' to the body; it is any point which has fixed location relative to the body-fixed axes (e.g., the centre of mass of a ring).

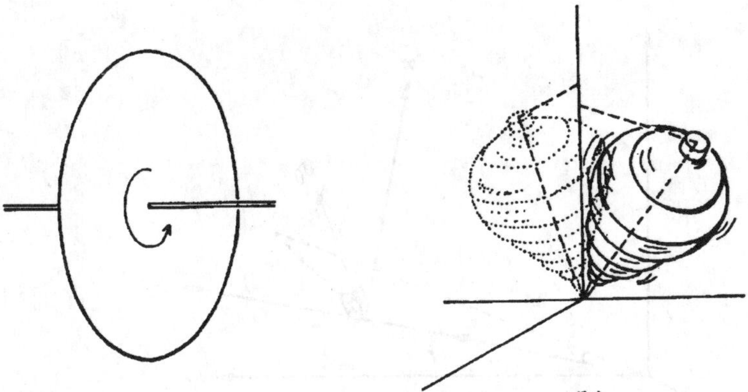

(a) (b)

Fig. 8.1 (a) A disc rotating about its axis. (b) A top rotating with one point fixed.

S5 *Can we not have intermediate cases: rotation about two or three fixed points, for example?*

T5 The axis of rotation must obviously pass through a fixed point. Since two points determine a line, this case is no different from rotation about a fixed axis. By the same token, rotation about three fixed points is impossible unless they are collinear.

Angular Velocity

S6 *I am not clear about the precise definition of angular velocity.*

T6 The notion of angular velocity can be introduced even for a point particle.

Suppose the position vector of a particle is $\boldsymbol{r}(t)$. Its displacement $\delta \boldsymbol{r}$ in time δt can be decomposed along the radial and the transverse directions:

$$\delta \boldsymbol{r} = \delta r \widehat{e}_r + r \delta\theta \widehat{e}_t,$$

where $\widehat{e}_r$ and $\widehat{e}_t$, respectively, are the radial and the transverse unit vectors in the plane defined by $\boldsymbol{r}$ and $\delta\boldsymbol{r}$, and $\delta\theta$ is the angular displacement in this plane (see Fig. 8.2). The angular velocity $\boldsymbol{\omega}$ is defined to be a vector whose magnitude is given by

$$|\boldsymbol{\omega}| = \lim_{\delta t \to 0} \frac{\delta\theta}{\delta t}$$

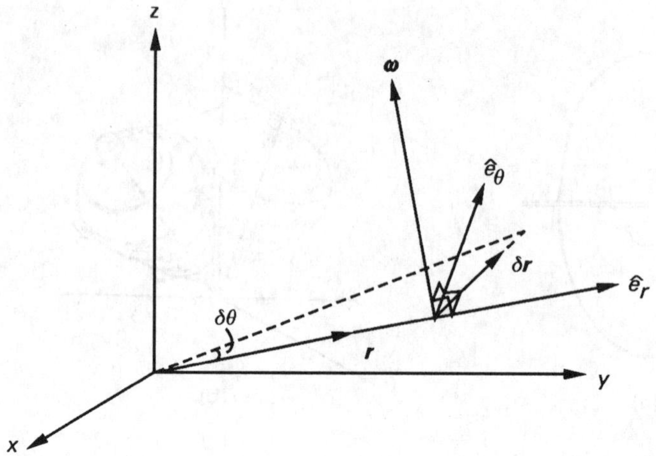

Fig. 8.2 $\delta\theta$ is in the plane of $\boldsymbol{r}$ and $\delta\boldsymbol{r}$. $\boldsymbol{\omega}$ is perpendicular to the plane.

and whose direction is perpendicular to both $\hat{e}_r$ and $\hat{e}_t$, such that $\boldsymbol{\omega}$, $\hat{e}_r$ and $\hat{e}_t$ form a right-handed triad. Thus

$$\delta\boldsymbol{r} = \delta r\hat{e}_r + \boldsymbol{\omega} \times \boldsymbol{r}\delta t.$$

Note, $\boldsymbol{\omega}$ for a point particle as defined here depends upon the origin. For the simple case of circular motion,

$$\delta\boldsymbol{r} = \boldsymbol{\omega} \times \boldsymbol{r}\delta t.$$

i.e. $\boldsymbol{v} = \boldsymbol{\omega} \times \boldsymbol{r}$

which checks with the elementary definition of angular velocity that you probably know.

S7 *What then does one mean by the angular velocity of a rigid body?*

T7 We have already introduced the notion of body-fixed frame. Relative to this frame the position vectors of all the particles of the rigid body are fixed. The angular velocity $\boldsymbol{\omega}$ of a rigid body is the angular velocity of its body-fixed frame. It is a vector which describes the rate of change of orientation of the body-fixed frame relative to the laboratory frame; its direction specifies the instantaneous axis of rotation of the body-fixed frame.

We saw in Chapter 5, how the rate of change of a vector relative to the laboratory frame is related to its rate of change relative to a rotating frame:

$$\left(\frac{d\boldsymbol{A}}{dt}\right)_{\text{laboratory}} = \left(\frac{d\boldsymbol{A}}{dt}\right)_{\text{body}} + \boldsymbol{\omega} \times \boldsymbol{A}.$$

Consider the motion of a rigid body with one point fixed in the laboratory frame. Take this point to be the origin of both the laboratory and the body-fixed frames. Let $\boldsymbol{r}_i$ be the position vector of the ith particle of the body. Using $\boldsymbol{r}_i$ for $\boldsymbol{A}$ in the above relation, and the fact that $(\delta\boldsymbol{r}_i)_{body} = 0$, we get

$$\delta\boldsymbol{r}_i = \boldsymbol{\omega} \times \boldsymbol{r}_i \delta t. \tag{2}$$

The important thing to note is that a single vector $\boldsymbol{\omega}$ specifies the infinitesimal displacements of all the particles of a rigid body. (This is of course related to the fact that pure rotation has three degrees of freedom.) Note that the infinitesimal displacement of every point of a rigid body is perpendicular to the angular velocity vector. Consequently, the infinitesimal displacements of various points lie in a set of parallel planes normal to $\boldsymbol{\omega}$.

We can easily see that Eq. (2) satisfies the rigid body constraints:

$$\delta r_i^2 = 2\delta\boldsymbol{r}_i \cdot \boldsymbol{r}_i = 0.$$

And, for any i and j,

$$\delta \boldsymbol{r}_{ij}^2 = 0,$$

where $\boldsymbol{r}_{ij} = \boldsymbol{r}_i - \boldsymbol{r}_j$.

S8 *What is meant by the 'instantaneous axis of rotation'?*

T8 In the body-fixed frame, the rigid body is at rest. From Eq. (2), in the laboratory frame the points which have $\boldsymbol{r}_i$ parallel to $\boldsymbol{\omega}$ are instantaneously at rest. These points lie on a line which passes through the fixed point and is parallel to $\boldsymbol{\omega}$. This line is the axis of rotation. The direction of $\boldsymbol{\omega}$ can vary with time, hence the name 'instantaneous axis of rotation'.

You must realize that $\boldsymbol{\omega}$ is a *free vector*. Since the axis of rotation passes through the fixed point and is parallel to $\boldsymbol{\omega}$, it is customary to draw $\boldsymbol{\omega}$ along the axis of rotation. It is wrong, however, to think of $\boldsymbol{\omega}$ as a vector localized at the fixed point. You will appreciate this as we go along.

S9 *What if no point of the rigid body is fixed?*

T9 As I have said in T3, the general motion of a rigid body is a rotation plus a translation. Let us understand this better.

Relative to the laboratory frame, various points of the body have different velocities. Choose any point of the body and imagine a frame (inertial) in which the chosen point is instantaneously at rest.† Relative to this frame, the motion of the rigid body is a pure rotation about the chosen point. This point in turn translates relative to the laboratory frame. Thus, the general motion of a rigid body can be viewed as a translation of an arbitrarily chosen point plus a pure rotation of the body in the instantaneous rest frame of the chosen point.

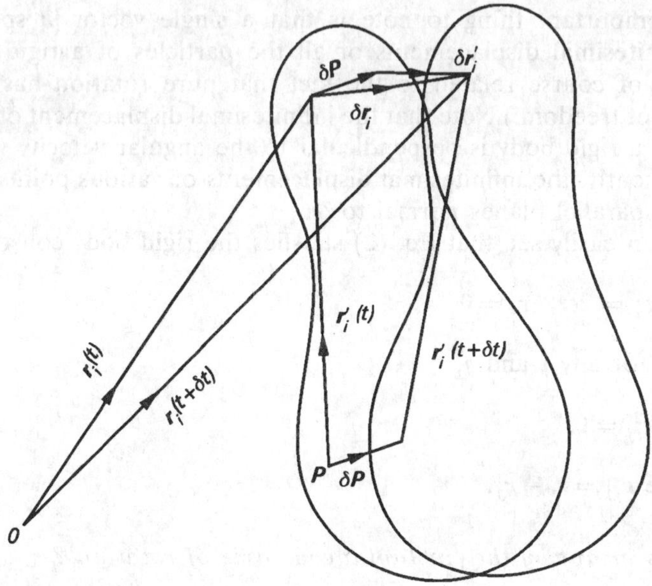

Fig. 8.3 An arbitrary infinitesimal displacement of a rigid body.

Let $\delta \boldsymbol{r}_i$ be the displacement (in time δt) of the ith particle of a rigid body relative to the laboratory frame. Let P be an arbitrary point of the rigid body and $\delta \boldsymbol{P}$ its displacement in the same time interval (See Fig. 8.3). Let $\delta \boldsymbol{r}_i'$ be the displacement of the particle relative to the instantaneous rest frame of P. (The origin of the instantaneous rest frame is at P.) If $\boldsymbol{\omega}$ is the angular velocity of the rigid body,

$$\begin{aligned} \delta \boldsymbol{r}_i &= \delta \boldsymbol{P} + \delta \boldsymbol{r}_i' \\ &= \delta \boldsymbol{P} + \boldsymbol{\omega} \times \boldsymbol{r}_i' \delta t \end{aligned} \tag{3}$$

Note that, while writing the second step, we have used Eq. (2) which is valid for rotation of a rigid body with one point fixed.

† This point, in general, will have an accelerated motion. Hence, an inertial frame can be the rest frame of such a point only instantaneously, i.e. a new inertial frame must be chosen at each instant.

Note also that the choice of the "fixed" point is not unique. We could just as well choose some other point Q of the body and view the general motion as a translational motion of Q plus a rotation about the "fixed" point Q.

S10 *That last remark is a little confusing. Can you give an example?*

T10 Consider a cylinder rolling down an inclined plane. This is an example of translational plus rotational motion. But rotation about which point? That is a matter of choice. As said earlier, the cylinder may be thought to be rotating about any point of the body. But two choices are particularly convenient. We can choose the CM of the cylinder as the "fixed" point. Its translational motion is rectilinear. The general motion, then, is the motion of the CM plus the rotation of the cylinder about the CM. As another choice we could consider the instantaneous point of contact as the "fixed" point. For perfect rolling, this point has no translational motion—its instantaneous velocity is zero. The motion of the cylinder is then viewed as a pure rotation about its instantaneous point of contact with the surface.

$\boldsymbol{\omega}$ has no reference to any origin

S11 *Does $\boldsymbol{\omega}$ not depend upon which point you consider to be "fixed"? I mean, depending upon whether you choose to look at the rigid body motion as a rotation about this point or that, will $\boldsymbol{\omega}$ not be different?*

T11 That is a natural question. We have introduced the angular velocity $\boldsymbol{\omega}$ of a rigid body via Eq. (2), which makes specific reference to the origin of the instantaneous rest frame (for the definition of $\boldsymbol{r}_i'$). But, as we shall see, this reference is spurious for the definition of $\boldsymbol{\omega}$. The angular velocity of a rigid body does not depend on which point you consider to be "fixed".

S12 *That seems a remarkable thing. How do you prove it?*

T12 Take two choices for the "fixed" point: P and Q as shown in Fig. 8.4. From the defining Eq. (3) of the angular velocity we have

$$\delta \boldsymbol{r}_i = \delta \boldsymbol{P} + \boldsymbol{\omega} \times \boldsymbol{r}_i' \delta t = \delta \boldsymbol{Q} + \boldsymbol{\omega}' \times \boldsymbol{r}_i'' \delta t$$

where $\boldsymbol{\omega}$ and $\boldsymbol{\omega}'$ are the angular velocities of the body about P and Q respectively. Now

$$\delta \boldsymbol{Q} = \delta \boldsymbol{P} + \boldsymbol{\omega} \times (\boldsymbol{r}_i' - \boldsymbol{r}_i'')\delta t.$$

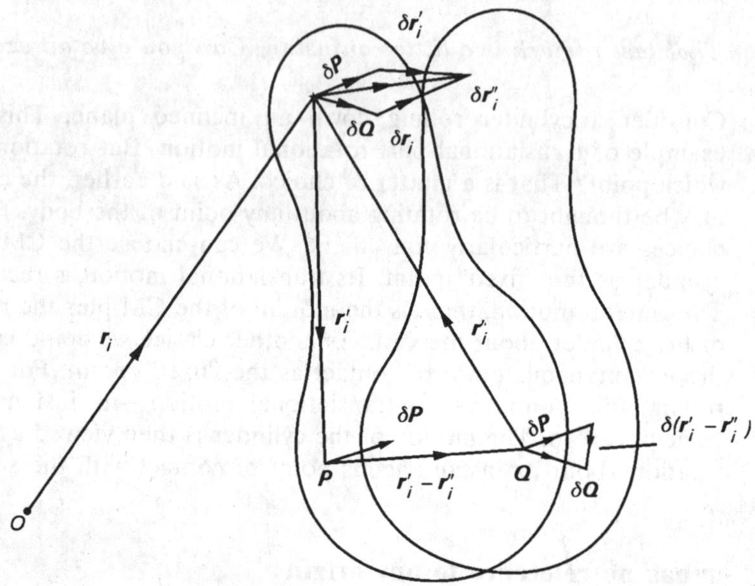

Fig. 8.4 An infinitesimal displacement of a rigid body viewed as a translation $\delta\boldsymbol{P}$ or $\delta\boldsymbol{Q}$ plus an appropriate rotation, $\delta\boldsymbol{r}_i'$ or $\delta\boldsymbol{r}_i''$ respectively.

These relations give

$$\boldsymbol{r}_i'' \times (\boldsymbol{\omega} - \boldsymbol{\omega}') = 0$$

Since $\boldsymbol{r}_i''$ is arbitrary, $\boldsymbol{\omega} = \boldsymbol{\omega}'$.†

The independence of $\boldsymbol{\omega}$ on the choice of the "fixed" point is also seen by looking at the change in the relative vector $\boldsymbol{r}_{ij}$:

$$\delta \boldsymbol{r}_{ij} = \boldsymbol{\omega} \times \boldsymbol{r}_{ij}' \delta t.$$

---

† Strictly, for a linear rigid body, $\boldsymbol{\omega}$ is undertermined to within addition of an arbitrary vector along the length of the body. This is evident from Eq. (2). For a linear rigid body, the component of $\boldsymbol{\omega}$ along the body has no physical significance; a linear rod cannot rotate about its length axis. Thus for a rod precessing about an axis the angular velocity is normal to the rod and changing its direction continuously. The indeterminacy of the component of angular velocity along the rod can however be exploited to construct a fixed angular velocity vector which is sometimes called the precessional angular velocity.

However, $\boldsymbol{r}'_{ij} = \boldsymbol{r}_{ij}$, hence,

$$\delta \boldsymbol{r}_{ij} = \boldsymbol{\omega} \times \boldsymbol{r}_{ij} \delta t. \tag{4}$$

Thus $\boldsymbol{\omega}$ defined here in terms of relative vectors needs no reference to any choice of origin.

S13 *I am clear that $\boldsymbol{\omega}$ is independent of the origin, but surely it will depend on the state of motion of the frame of reference. For example relative to the body-fixed frame, $\boldsymbol{\omega}$ must be zero.*

T13 That is a common error, and you will appreciate it if you consider the analogous situation in the translational case. Consider a body moving with a velocity $\boldsymbol{v}$ relative to the ground's frame. Let its velocity, relative to a train moving with velocity $\boldsymbol{u}$, be $\boldsymbol{v}'$. Clearly $\boldsymbol{v} = \boldsymbol{u} + \boldsymbol{v}'$. Now in order to find velocity of the train relative to train's frame, we substitute $\boldsymbol{u}$ for $\boldsymbol{v}$, which gives $\boldsymbol{v}' = 0$ as expected. But it will be absurd to say that $\boldsymbol{u}$ is zero in the train's frame since $\boldsymbol{u}$ by definition stands for the velocity of the train relative to the ground's frame.

Similarly $\boldsymbol{\omega}$ by definition, is the angular velocity of the body relative to the laboratory frame, and it is silly to say that $\boldsymbol{\omega}$ is zero in the rest frame of the rigid body. True, that relative to its rest frame, the angular velocity of the rigid body is zero, but that is not what $\boldsymbol{\omega}$ stands for.

S14 *Then is it correct to say that $\boldsymbol{\omega}$ is the same in all frames?*

T14 Make sure you are clear about what you mean here. $\boldsymbol{\omega}$ is a definite vector: the angular velocity of the rigid body relative to the laboratory frame. Relative to different frames it has different components; but it is a vector that relates two given frames (the body-fixed and the laboratory frames) and it is meaningless to talk of its transformation from one frame to another.

If, however, you mean to ask whether the angular velocity of the rigid body is the same relative to all frames, then that is a different matter. Equation (4) is the defining relation for the angular velocity of a rigid body relative to the (inertial) laboratory frame. We may rewrite it as

$$\boldsymbol{v}_{ij} = \boldsymbol{\omega} \times \boldsymbol{r}_{ij}. \tag{5}$$

Now consider another inertial frame moving with a velocity $\boldsymbol{v}$ relative to the laboratory frame. We have

$$\delta \boldsymbol{r}'_i = \delta \boldsymbol{r}_i - \boldsymbol{v} \delta t,$$

and a similar relation for $\delta \boldsymbol{r}_j$, which means $\delta \boldsymbol{r}'_{ij} = \delta \boldsymbol{r}_{ij}$. Thus

$$\boldsymbol{r}'_{ij} = \boldsymbol{r}_{ij} \quad \text{and} \quad \boldsymbol{v}'_{ij} = \boldsymbol{v}_{ij}. \tag{6}$$

Equations (5) and (6) show that the angular velocity of the rigid body relative to any inertial frame is the same as that relative to the laboratory frame, viz. $\boldsymbol{\omega}$.

Next, consider a rotating (non-inertial) frame with an angular velocity ω_0 relative to an inertial frame. If $(\boldsymbol{v}_{ij})_{\text{rot}}$ is the rate of change of the relative vector $\boldsymbol{r}_{ij}$ with respect to this frame, we have

$$\boldsymbol{v}_{ij} = (\boldsymbol{v}_{ij})_{\text{rot}} + \boldsymbol{\omega}_0 \times \boldsymbol{r}_{ij},$$

which, using Eq. (5), gives

$$(\boldsymbol{v}_{ij})_{\text{rot}} = (\boldsymbol{\omega} - \boldsymbol{\omega}_0) \times \boldsymbol{r}_{ij}.$$

Thus, relative to the rotating frame, the angular velocity of the body is $(\boldsymbol{\omega} - \boldsymbol{\omega}_0)$. In particular, the angular velocity of the rigid body relative to its body-fixed frame is zero.

Note, however, that the meaning of $\boldsymbol{\omega}$ is still the same: it is the angular velocity of the rigid body relative to the laboratory frame (or any other inertial frame). It is <u>*wrong*</u> to say that $\boldsymbol{\omega}$ is zero relative to the body-fixed frame.

S15 *How to ascertain the axis of rotation when motion involves translation and rotation?*

T15 All points with $\boldsymbol{r}'_i$ parallel to $\boldsymbol{\omega}$ have $\delta \boldsymbol{r}_i = \delta \boldsymbol{P}$ [Eq. (3)], i.e. they have the same translational motion relative to the laboratory frame as the chosen point. The locus of these points is the axis of rotation. Since, as we have seen, the choice of the "fixed" point P is arbitrary the axis of rotation is not unique. Its direction, however, is unique, being parallel to $\boldsymbol{\omega}$.

S16 *Can we consider some examples to consolidate what we have discussed above?*

T16 Let us start with the simplest example: the rotation of a flywheel about the axis passing through its centre and perpendicular to its plane. This is a special case of rotation about a fixed point—the axis of rotation is also fixed. We can, however, consider another point P on the flywheel and view the motion as rotation about P provided we also add to it the translational motion (in this case, circular motion) of the point P. This would, of course be a silly thing to do, but it is not wrong and the important thing is that the angular velocity of the wheel about P is the same as that about the centre. This is what we mean by saying that $\boldsymbol{\omega}$ is not a localized vector.

Consider next a rolling cylinder. What is its axis of rotation? To answer that we appeal to the defining relation:

$$\delta \boldsymbol{r}_{ij} = \boldsymbol{\omega} \times \boldsymbol{r}_{ij} \delta t.$$

Clearly, $\delta \boldsymbol{r}_{ij} = 0$ for $\boldsymbol{r}_{ij}$ parallel to $\boldsymbol{\omega}$. Now, the relative vectors $\boldsymbol{r}_{ij}$ which are displaced parallel to themselves have $\delta \boldsymbol{r}_{ij} = 0$, and hence are parallel to $\boldsymbol{\omega}$. For a rolling cylinder any line parallel to the cylinder axis is displaced parallel to itself. Therefore relative vectors parallel to the cylinder axis do not change during the rolling motion. Consequently, the angular velocity vector $\boldsymbol{\omega}$ of the cylinder is parallel to the cylinder axis. This is an example of translational-cum-rotational motion, with the simplifying feature that the direction of $\boldsymbol{\omega}$ is fixed.

As stressed repeatedly $\boldsymbol{\omega}$ has no reference to any origin. We could consider the CM or the instantaneous point of contact or any other point as the "fixed" point about which the cylinder rotates. The translational motion is different in each case and so is the axis of rotation, *but $\boldsymbol{\omega}$ is the same*. This crucial point is rarely emphasized (see Box 8.1).

Finally consider the motion of a top with one point fixed. This is an example of rotation about a fixed point but with continuously changing $\boldsymbol{\omega}$. The details are hard to handle here and can be looked up in any standard textbook. The instantaneous $\boldsymbol{\omega}$ of the top is once again given by Eq. (4).

As described in T4, as the top spins about its symmetry axis, it also precesses about a vertical axis and nutates. Since the spinning motion is the most conspicuous, one often tends to equate the axis of rotation to the symmetry axis. This, however, is only approximately true. A typical point of the top, in an infinitesimal interval of time, moves around the axis of symmetry (spin). It also proceeds along a small distance in the horizontal direction (precession) and moves slightly up or down (nutation). Consequently, the infinitesimal displacements of the points of a top define planes which are not quite perpendicular to its symmetry axis. The angular velocity of a top, therefore, is not quite along its axis of symmetry. The spin motion of a top is the fastest, and hence the component of angular velocity along the symmetry axis is the largest. A small component, transverse to the symmetry axis, describes the precessional and nutational angular velocity of the top.

Box 8.1

A rolling disc

a) Let the CM of the disc be the chosen "fixed" point. The displacement $\delta \boldsymbol{r}$ of a point P in time δt relative to the laboratory frame is given by $(\boldsymbol{v}_{\text{CM}} + \boldsymbol{\omega} \times \boldsymbol{r}')\delta t$ as shown in Fig. 8.5(a). For the CM itself $\boldsymbol{r}' = 0$, so that its displacement is $\boldsymbol{v}_{\text{CM}}\,\delta t$. The instantaneous point of contact A has $r'_A = R$. Clearly, $\boldsymbol{\omega} \times \boldsymbol{r}'_A = -\boldsymbol{v}_{\text{CM}}$ for A since its displacement relative to the laboratory frame vanishes (rolling condition). For point B diametrically opposite to A, $\boldsymbol{\omega} \times \boldsymbol{r}'_B = \boldsymbol{v}_{\text{CM}}$ and its velocity relative to the laboratory frame is thus twice that of the CM.

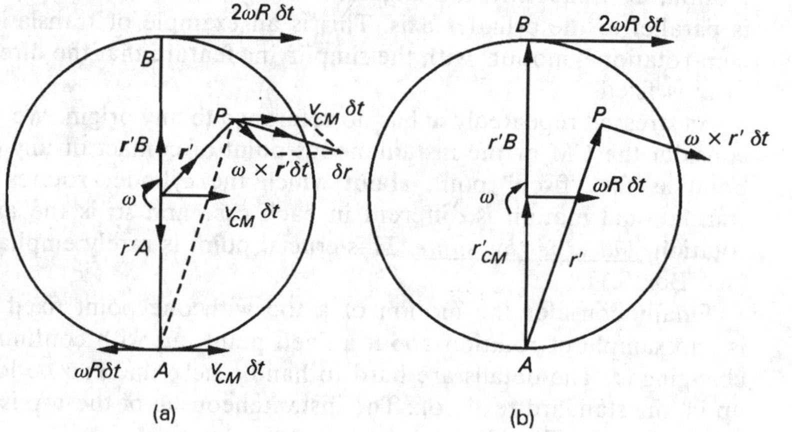

Fig. 8.5 A rolling cylinder. (a) The CM is the chosen "fixed" point. (b) The instantaneous point of contact is the chosen "fixed" point.

b) Let the instantaneous point of contact A be the "fixed" point. For rolling, the instantaneous velocity of A relative to the laboratory frame is zero. Therefore the displacement of a point at $\boldsymbol{r}'$, in time δt, relative to the laboratory frame is given by $\boldsymbol{\omega}' \times \boldsymbol{r}'\ \delta t$ as shown in Fig. 8.5(b). *Note that $\boldsymbol{r}'$ is now relative to A.* The displacement of the CM is $\boldsymbol{\omega}' R\delta t$. Clearly this must equal $v_{\text{CM}}\,\delta t$ as in (a), since displacement of a point relative to a given frame cannot depend on the choice of the "fixed" point. It thus follows that $\boldsymbol{\omega}' = \boldsymbol{\omega}$. For point B, $r'_B = 2R$ and the velocity is $2R\boldsymbol{\omega}$ as before.

S17 *The angular velocity of a particle that you talked about in T6 depends upon the origin whereas that of a rigid body does not. How do you understand this?*

T17 A point does not have size. There is no relative vector $\boldsymbol{r}_{ij}$ and the angular velocity, in the sense of Eq.(4), is undefined for a point object.

The general motion of a rigid body is a translation plus a rotation, whereas the motion of a point object is always purely translational. The angular velocity of a rigid body is associated with its rotational degrees of freedom. The angular velocity of a point object that we defined in T6 is associated with the translational degrees of freedom of a point object.

We may resolve, as in T6, the translational displacement of a rigid body (i.e. displacement $\delta\boldsymbol{P}$ of the chosen point) into transverse and radial components, and introduce an angular velocity for its translational motion. This angular velocity, then, will depend on the origin just as that of a point object. But, clearly, this is not what we mean by the angular velocity of the rigid body.

S18 *You said the possible motion of a point object is only translational. But how about circular motion? Is it not pure rotation of a point object?*

T18 A pure rotation requires, by definition, that at least one point associated with an object be fixed. So the motion of a point object, performing any kind of motion, is necessarily translational. A point object has no rotational degrees of freedom. Circular motion is simply a special case of translational motion where the radial velocity relative to its centre is zero.

Moment of Inertia

S19 *Let us turn to angular momentum. How is it related to angular velocity?*

T19 Consider the motion of a rigid body with one point fixed in the laboratory frame. The angular momentum of the body in the laboratory frame about the fixed point is

$$\boldsymbol{L} = \sum_{\ell=1}^{N} m_\ell \boldsymbol{r}_\ell \times \boldsymbol{v}_\ell .$$

For a rigid body, we have

$$\boldsymbol{v}_\ell = \boldsymbol{\omega} \times \boldsymbol{r}_\ell ,$$

which gives, after expanding the vector triple product,

$$\mathbf{L} = \sum_{\ell=1}^{N} m_\ell [r_\ell^2 \boldsymbol{\omega} - (\mathbf{r}_\ell \cdot \boldsymbol{\omega})\mathbf{r}_\ell].$$

Since $\boldsymbol{\omega}$ cannot be factored out on the right-hand side, let us write this equation in component form

$$L_i = \sum_{\ell=1}^{N} m_\ell [r_\ell^2 \omega_i - \left(\sum_j (r_\ell)_j \omega_j\right)(r_\ell)_i] \\ = \sum_j I_{ij}\omega_j, \tag{7}$$

$$\text{where} \quad I_{ij} = \sum_{\ell=1}^{N} m_\ell [r_\ell^2 \delta_{ij} - (r_\ell)_i (r_\ell)_j] \quad i \cdot j = x, y, z. \tag{8}$$

I_{ij} is called the moment of inertia tensor of the body. Let us display its components in matrix form

$$\{I_{ij}\} = \begin{pmatrix} \sum m_\ell(r_\ell^2 - x_\ell x_\ell) & \sum m_\ell(-x_\ell y_\ell) & \sum m_\ell(-x_\ell z_\ell) \\ \sum m_\ell(-y_\ell x_\ell) & \sum m_\ell(r_\ell^2 - y_\ell y_\ell) & \sum m_\ell(-y_\ell z_\ell) \\ \sum m_\ell(-z_\ell x_\ell) & \sum m_\ell(-z_\ell y_\ell) & \sum m_\ell(r_\ell^2 - z_\ell z_\ell) \end{pmatrix} \tag{9}$$

In the dyadic notation, Eq. (7) can be expressed as

$$\mathbf{L} = \overleftrightarrow{\mathbf{I}} \cdot \boldsymbol{\omega}, \tag{10}$$

where the moment of inertia dyadic is given by

$$\overleftrightarrow{\mathbf{I}} = \sum_{\ell=1}^{N} m_\ell [r_\ell^2 \mathbf{I} - \mathbf{r}_\ell \mathbf{r}_\ell] \tag{11}$$

with $\mathbf{I} = \widehat{ii} + \widehat{jj} + \widehat{kk}$ being the unit dyadic.

S20 *The linear momentum and velocity vectors are always parallel to each other. Is the same true for angular momentum and angular velocity vectors?*

T20 No, in general the angular momentum is not in the same direction as the angular velocity of a body. The quantity (moment of inertia) that relates the two is, therefore, a tensor of rank two.† (In contrast, the quantity (mass) relating the linear momentum and velocity is a scalar.) For an example of where $\boldsymbol{L}$ is not parallel to $\boldsymbol{\omega}$, see p.19 and p.20, Ch.9.

S21 *But we so often use the simple relation $\mathbf{L} = I\boldsymbol{\omega}$ in solving problems in mechanics. Here $\mathbf{L}$ and $\boldsymbol{\omega}$ are certainly parallel. What is the justification?*

T21 Your question requires discussion in some detail. Let us understand this quantity I_{ij} that we have called the moment of inertia tensor of a body.

As its definition [Eq. (8)] shows, the moment of inertia of a body is determined by the mass distribution of the body relative to the chosen co-ordinate system. With respect to body-fixed axes, it is a purely geometric property. Further, I_{ij} is a symmetric second rank tensor [see Eq. (9)), so it has only six independent components.

The values of the components, obviously, depend on the origin and orientation of the body-fixed axes relative to which they are calculated. There is an important result of mechanics which asserts that *for any rigid body there exists a body-fixed orthogonal co-ordinate system relative to which the moment of inertia tensor is diagonal.* These special axes are called the principal axes of the rigid body and the three diagonal components are called the principal moments of inertia.

S22 *How does one locate the principal axes of a rigid body?*

T22 There is a standard method of locating the principal axes of a rigid body and obtaining its principal moments of inertia. This involves solving the so-called eigenvalue equation of the matrix I_{ij}. This, however, is beyond our present scope. Nevertheless, I can give you a result, without proof, which is very useful. *Every symmetry axis of a body is a possible principal axis.*

For example, a sphere (with uniform mass distribution) is symmetrical about any of its diameters. So take any three mutually perpendicular diameters and they form a set of principal axes. A cube is symmetrical about the three mutually perpendicular lines which pass through its centre and intersect its faces in their centres. So these three lines are a set of principal axes of the cube. The major axis and any two mutually perpendicular minor axes form a set of principal axes for an ellipsoid. Figure 8.6 shows many common shapes and the corresponding principal axes.

† See Pages 34–36, Ch. 2.

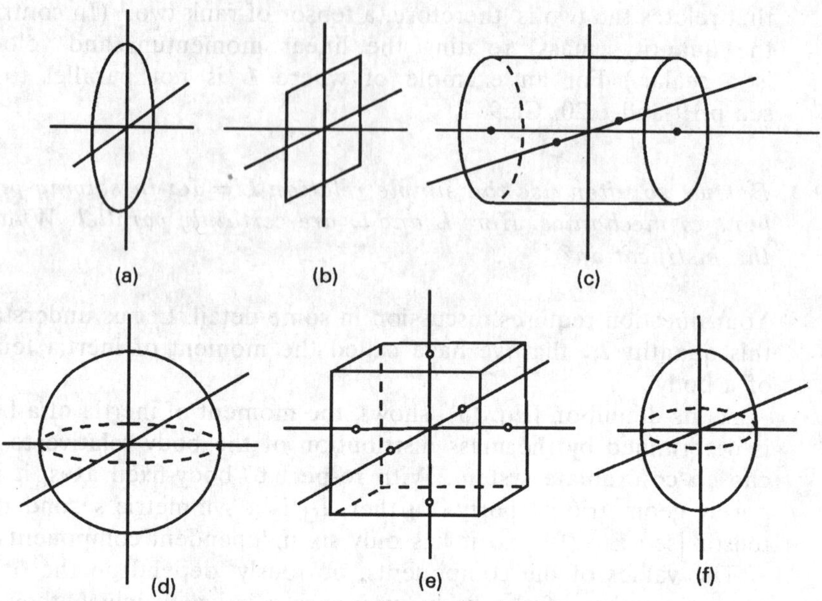

Fig. 8.6 The principal axes of objects of various shapes. (a) Disc (b) Rectangular lamina (c) Cylinder (d) Sphere (e) Cube (f) Ellipsoid

S23 *In what way does the choice of principal axes help?*

T23 The principal axes are obviously convenient to work with. Instead of six components we have only three diagonal components of I_{ij} and Eq. (7) simplifies to $L_x = I_{xx}\omega_x, L_y = I_{yy}\omega_y$ and $L_z = I_{zz}\omega_z$. Note that, since I_{xx}, I_{yy} and I_{zz} may be different, $\boldsymbol{L}$ and $\boldsymbol{\omega}$ are, in general, not parallel.

Now if $\boldsymbol{\omega}$ is along one of the principal axes, two of its components are zero and, consequently, the corresponding components of $\boldsymbol{L}$ are also zero. For example, if $\boldsymbol{\omega}$ is along z-axis, $\omega_x = \omega_y = 0$ and hence $L_x = L_y = 0$ while $L_z = I_{zz}\omega_z$. This last relation can be written as $\boldsymbol{L} = I\boldsymbol{\omega}$.

In elementary physics we mostly deal with such situations. For example a fly-wheel rotating about its axis is rotating with $\boldsymbol{\omega}$ parallel to a principal axis. A sphere rotating about one of its diameters or a cylinder rotating about its axis also have $\boldsymbol{\omega}$ parallel to one of their principal axes.

Another situation where $\boldsymbol{L} = I\boldsymbol{\omega}$ holds is when all the principal moments are equal, i.e. $I_{xx} = l_{yy} = I_{zz}$. This is true, for example, for the symmetry axes of a sphere or a cube. Note, in this situation, there is no restriction on the direction of $\boldsymbol{\omega}$.†

† This means $\boldsymbol{L} = I\boldsymbol{\omega}$ is, in general, true for such bodies. This, in turn, implies that

S24 *Incidently, why is I_{ij} called the 'moment of inertia' tensor ?*

T24 In physics, moments of a distribution $\rho(\boldsymbol{r})$, say mass distribution or charge distribution, relative to a co-ordinate system involve successive products of $\boldsymbol{r}_i, \boldsymbol{r}_i\boldsymbol{r}_j$ etc. with $\rho(\boldsymbol{r})$ integrated over space. For example, the electric dipole moment, quadrupole moment are quantities defined in this manner for charge density $\rho(\boldsymbol{r})$. I_{ij}, in this language, is a moment of the inertial mass density. Hence the name 'moment of inertia'. The word inertia is appropriate on another count also. Just as mass m is a measure of 'inertia' for translational motion, I is a measure of 'inertia' in rotational motion. This is evident from the rotational law of motion

$$I\frac{d\boldsymbol{\omega}}{dt} = \boldsymbol{N},$$

which is familiar to you from elementary physics.

Euler's Equations of Motion

S25 *From your preceding discussion I anticipate that this form will not hold in general. What is the general law of rotational motion of a rigid body?*

T25 If the body has one point fixed in the laboratory frame, the general law of rotational motion is

$$\left(\frac{d\boldsymbol{L}}{dt}\right)^{(\text{laboratory})} = \boldsymbol{N}^{\text{ext}}.$$

The superscript 'laboratory' has been added to remind us that the quantities are to be evaluated relative to the laboratory frame about the fixed point as the origin.† Using Eq. (10) we may rewrite the equation as

$$\left(\frac{d}{dt}\overleftrightarrow{\boldsymbol{I}}\cdot\boldsymbol{\omega}\right)^{(\text{laboratory})} = \boldsymbol{N}^{\text{ext}}.$$

any three independent axes of such a body with origin at its centre are principal axes, with equal diagonal elements for the moment of inertia tensor with respect to them.

† For the general case of translation plus rotation, the convenient reference point is the CM of the rigid body. We have seen in T11, Ch. 6 that the same law as in Eq. (1) holds in the CM frame even though the CM frame may be accelerating and hence non-inertial. Further, the translational motion of the CM is governed by the simple law in Eq. (1).

Now, during the course of the motion, the distribution of the mass of the rigid body about a co-ordinate system fixed in the laboratory frame goes on changing. Clearly, relative to the laboratory frame, the moment of inertia is time-dependent and cannot be taken out of the time differentiation. To do that we must turn to the body-fixed frame with the same origin and use the relation

$$\left(\frac{d\boldsymbol{L}}{dt}\right)^{\text{(laboratory)}} = \left(\frac{d\boldsymbol{L}}{dt}\right)^{\text{(body)}} + \boldsymbol{\omega} \times \boldsymbol{L},$$

which gives

$$\left(\frac{d\boldsymbol{L}}{dt}\right)^{\text{(body)}} + \boldsymbol{\omega} \times \boldsymbol{L} = \boldsymbol{N}^{\text{ext}}.$$

We thus obtain

$$\left(\frac{d\,\overleftrightarrow{\boldsymbol{I}} \cdot \boldsymbol{\omega}}{dt}\right)^{\text{(body)}} + \boldsymbol{\omega} \times \boldsymbol{L} = \boldsymbol{N}^{\text{ext}}.$$

Now, the moment of inertia relative to the body-frame is constant in time. Further, we choose the principal axes for the co-ordinate axes of the body-frame so that I_{ij} is diagonal. The above equation in component form then reduces to:

$$I_{xx}\frac{d\omega_x}{dt} - \omega_y\omega_z(I_{yy} - I_{zz}) = N_x,$$

$$I_{yy}\frac{d\omega_y}{dt} - \omega_z\omega_x(I_{zz} - I_{xx}) = N_y,$$

and $$I_{zz}\frac{d\omega_z}{dt} - \omega_x\omega_y(I_{xx} - I_{yy}) = N_z.$$

These are the well-known Euler's equations of rigid body rotation. Although we have dropped the superscript 'body', we must remember that the components in this relation refer to the principal axes of the body-frame. $\boldsymbol{\omega}$ however, is the angular velocity of the rigid body relative to the laboratory frame as emphasized in T13. The time derivative of $\boldsymbol{\omega}$ may be considered relative to the body-frame or space-frame, since the equation in T7 relating the rates of change of a vector relative to the two frames gives

$$\left(\frac{d\boldsymbol{\omega}}{dt}\right)^{\text{space}} = \left(\frac{d\boldsymbol{\omega}}{dt}\right)^{\text{body}}$$

S26 *In which situations do we retrieve the simple equation $I\alpha = N^{ext}$?*

T26 These are the same situations for which $\boldsymbol{L} = I\boldsymbol{\omega}$ holds as explained in T23: either the body has equal principal moments of inertia or, for a general body, $\boldsymbol{\omega}$ is along one of its principal axes. For example you can apply this relation for the angular acceleration due to the torque on a disc rotating about its axis or to a sphere or a cube rotating about any axis passing through its centre.

S27 *An aside. A disc rotating about its axis with a constant angular velocity has no external forces or torques. Yet every element of it must need a centripetal force to keep moving on a circle. What is the agency of this force?*

T27 Your observation is correct. Every element of the disc indeed has a centripetal acceleration and does experience a centripetal force. These forces are provided by rigid body constraints. You see, the various elements of a rigid body are constrained to be at definite distances from one another in a definite configuration. Any tendency of deviation from this configuration produces (infinitesimal) radial strains in the body, as a result of which (finite) radial stresses are produced. These stresses provide the necessary centripetal force for the rotational motion of the body.

S28 *What is the essential feature of rotation about a point that is different from rotation about a fixed axis?*

T28 In the case of rotation about a fixed axis, which is a principal axis of the body, $\boldsymbol{L} = I\boldsymbol{\omega}$. The direction of $\boldsymbol{L}$ is fixed; only its magnitude can change.

When the fixed axis is not a principal axis, $\boldsymbol{L}$ and $\boldsymbol{\omega}$ are not parallel and $\boldsymbol{L}$ can change both in magnitude as well as direction. This follows from $L_i = \sum I_{ij}\omega_j$, where the components are taken with respect to the laboratory axes. Though, for a fixed axis of rotation, ω_j are constant, I_{ij} are not. Hence, $\boldsymbol{L}$ can change with time.

In the case of rotation about a point, $\boldsymbol{\omega}$ can change both in its magnitude and direction and so can the angular momentum. You see an example of this in the precessional motion of a top.

S29 *Coming to the motion of a top, I have always wondered why the top does not simply topple under its weight.*

T29 Let us first understand why a planet revolving round the sun does not fall into the sun due to the gravitational attraction. You see, the direction of motion of a body is determined not only by the force acting on the body but also by its initial velocity. The force determines the change in velocity, which is in the direction of the force. But this gets added to the initial velocity to produce the final velocity of the body. The final velocity, therefore, may or may not be in the direction of the force.

In the case of the motion of a planet, at every instant, the velocity of the planet points almost perpendicular to the line joining the planet and the sun. The gravitational force, pointing towards the sun brings about a centripetal change in the velocity. This change added to the instantaneous velocity produces the new velocity, which is perpendicular to the new position of the line joining the planet and the sun.

Now consider a top spinning with its axis making an angle with the vertical. The instantaneous angular momentum of the top points almost along the axis of the top.† The weight of the top acts at its centre of mass in a vertically downward direction. The torque of the weight, about the point of contact with the ground, is obviously perpendicular to the plane containing the axis of the top and the line of action of the weight (see Fig. 8.7). According to the rotational law of motion the change in angular momentum must also point in the same direction. This change displaces the angular momentum sideward resulting in the precession of the top.

This is a rather crude description. Detailed dynamics of a top is quite involved for which you will need to consult an advanced text-book.

S30 *The centre of mass of the top, while precessing, seems accelerated; it moves along a circle. What provides the centripetal force for this motion? The weight obviously cannot do that; its direction is wrong.*

T30 The only forces acting on the top are: its weight and the reaction of the ground acting at the point of contact. Ignoring nutation, the centre of mass moves in a horizontal circle; there is no acceleration of the centre of mass in the vertical direction. So the net force along the vertical direction must be zero. The vertical reaction force at the point of contact must, therefore, be equal to the weight of the top. It is the horizontal component of the contact force of the ground that accounts for the centripetal acceleration of the centre of mass of the top. This component can arise from friction or any other suitable constraint force.

† The top, besides spinning, precesses and nutates. (See T4.) However, the angular velocities for the latter two modes of motion are usually much smaller compared to the spin. Hence, $\boldsymbol{L}$ is almost along the axis of the top.

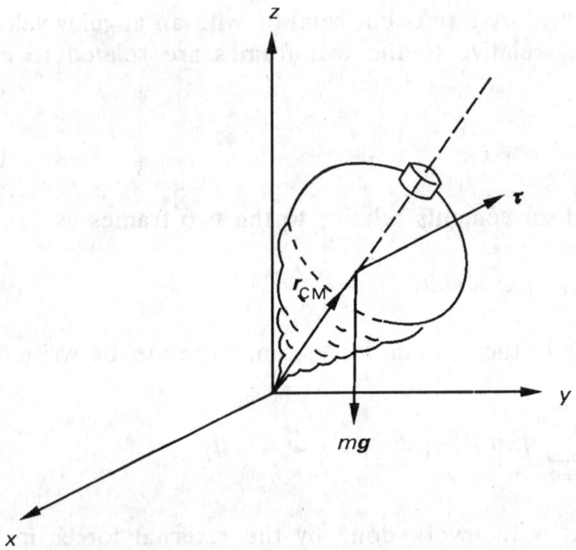

Fig. 8.7 The torque τ, and hence δL, is perpendicular to the plane containing r_{CM} and mg.

Work–energy theorem in a rotating frame

S31 *I want to return to the topic of the work–energy theorem. What does it look like in a rotating frame?*

T31 We have seen that the work–energy theorem for a system of N particles can be written as

$$\Delta W^{ext} + \Delta W^{int} = \Delta T. \tag{12}$$

where $$\Delta W^{ext} = \sum_{i=1}^{N} \int_1^2 \boldsymbol{F}_i^{ext} \cdot d\boldsymbol{r}_i \tag{13}$$

is the work done by the external forces and

$$\Delta W^{int} = \sum_{i=1}^{N} \int_1^2 \boldsymbol{F}_i^{int} \cdot d\boldsymbol{r}_i \tag{14}$$

is the work done by the internal forces. ΔT is the change in the kinetic energy of the system. All the quantities in this expression refer to the laboratory frame. Now consider a frame with the same origin as the laboratory frame, but rotating with an angular velocity $\boldsymbol{\omega}$. The velocities relative to the two frames are related to each other as

$$\boldsymbol{v}_i = (\boldsymbol{v}_i)_{\text{rot}} + \boldsymbol{\omega} \times \boldsymbol{r}_i \tag{15}$$

and hence, the displacements relative to the two frames as

$$d\boldsymbol{r}_i = (d\boldsymbol{r}_i)_{\text{rot}} + \boldsymbol{\omega} \times \boldsymbol{r}_i dt.$$

The work done by the external forces can, therefore, be written as

$$\Delta W^{\text{ext}} = \sum_{i=1}^{N} \int_1^2 \boldsymbol{F}_i^{\text{ext}} \cdot [(d\boldsymbol{r}_t)_{\text{rot}} + \boldsymbol{\omega} \times \boldsymbol{r}_i dt].$$

The first term is the work done by the external forces in the rotating frame

$$(\Delta W^{\text{ext}})_{\text{rot}} = \sum_{i=1}^{N} \int_1^2 \boldsymbol{F}_i^{\text{ext}} \cdot (d\boldsymbol{r}_i)_{\text{rot}}. \tag{16}$$

The second term can be rearranged as follows:

$$\sum_{i=1}^{N} \int_1^2 \boldsymbol{F}_i^{\text{ext}} \cdot \boldsymbol{\omega} \times \boldsymbol{r}_i dt = \int_1^2 \left(\sum_{i=1}^{N} \boldsymbol{r}_i \times \boldsymbol{F}_i^{ext} \right) \cdot \boldsymbol{\omega} dt.$$

The quantity in the parenthesis is the total external torque on the system, $\boldsymbol{N}^{\text{ext}}$, and $\boldsymbol{\omega} dt$ is the angular displacement $d\boldsymbol{\theta}$ of the rotating frame in time dt. Thus the work done by the external forces may be written as

$$\Delta W^{\text{ext}} = (\Delta W^{\text{ext}})_{\text{rot}} + \textstyle\int_1^2 \boldsymbol{N}^{\text{ext}} \cdot d\boldsymbol{\theta}. \tag{17}$$

Similarly the work done by the internal forces can be written as

$$\Delta W^{\text{int}} = (\Delta W^{\text{int}})_{\text{rot}} + \int_1^2 \boldsymbol{N}^{\text{int}} \cdot d\boldsymbol{\theta}.$$

If the internal forces satisfy the III law in the strong form, the total internal torque is zero and

$$\Delta W^{\text{int}} = \Delta W^{\text{int}}_{\text{rot}}. \tag{18}$$

The change in kinetic energy can be expressed, using Eq. (15), as

$$\begin{aligned}\Delta T &= \Delta \sum_{i=1}^{N} \frac{1}{2} m_i v_i^2 \\ &= \Delta \sum_{i=1}^{N} \frac{1}{2} m_i [(\boldsymbol{v}_i)_{\text{rot}} + \boldsymbol{\omega} \times \boldsymbol{r}_i]^2 \\ &= \Delta \sum_{i=1}^{N} \frac{1}{2} m_i (\boldsymbol{v}_i)^2_{\text{rot}} + \Delta \sum_{i=1}^{N} \frac{1}{2} m_i (\boldsymbol{\omega} \times \boldsymbol{r}_i)^2 \\ &\quad + \Delta \sum_{i=1}^{N} m_i (\boldsymbol{v}_i)_{\text{rot}} \cdot \boldsymbol{\omega} \times \boldsymbol{r}_i .\end{aligned}$$

The first term is the kinetic energy of the system in the rotating frame. The second term can be rearranged as follows:

$$\begin{aligned}\Delta \sum_{i=1}^{N} \frac{1}{2} m_i (\boldsymbol{\omega} \times \boldsymbol{r}_i)^2 &= \Delta \sum_{i=1}^{N} \frac{1}{2} m_i (\boldsymbol{\omega} \times \boldsymbol{r}_i) \cdot (\boldsymbol{\omega} \times \boldsymbol{r}_i) \\ &= \Delta \frac{1}{2} \boldsymbol{\omega} \cdot \sum_{i=1}^{N} m_i \boldsymbol{r}_i \times (\boldsymbol{\omega} \times \boldsymbol{r}_i) \\ &= \Delta \frac{1}{2} \boldsymbol{\omega} \cdot \overleftrightarrow{I} \cdot \boldsymbol{\omega} .\end{aligned}$$

where we have used the definition of the moment of inertia dyadic given in Eq. (11).

The last term can be rewritten in the form:

$$\begin{aligned}\Delta \sum_{i=1}^{N} m_i (\boldsymbol{v}_i)_{\text{rot}} \cdot (\boldsymbol{\omega} \times \boldsymbol{r}_i) &= \Delta \boldsymbol{\omega} \cdot \sum_{i=1}^{N} \boldsymbol{r}_i \times m_i (\boldsymbol{v}_i)_{\text{rot}} , \\ &= \Delta \boldsymbol{\omega} \cdot (\boldsymbol{L})_{\text{rot}}\end{aligned}$$

where $(\boldsymbol{L})_{\text{rot}}$ is the total angular momentum of the system relative to the rotating frame. We may thus, finally write

$$\begin{aligned}&(\Delta W^{\text{ext}})_{\text{rot}} + \int_1^2 \boldsymbol{N}^{\text{ext}}_{\text{tot}} \cdot d\boldsymbol{\theta} + (\Delta W^{\text{int}})_{\text{rot}} \\ &= \Delta \sum_{i=1}^{N} \frac{1}{2} m_i (\boldsymbol{v}_i)^2_{\text{rot}} + \Delta \frac{1}{2} \boldsymbol{\omega} \cdot \overleftrightarrow{I} \cdot \boldsymbol{\omega} + \Delta \boldsymbol{\omega} \cdot (\boldsymbol{L})_{\text{rot}} . \qquad (19)\end{aligned}$$

This is the work–energy theorem for a system of N particles in a frame rotating with angular velocity $\boldsymbol{\omega}$.

If the system of particles happens to be a rigid body, and the rotating frame is the body-frame, then all the particles are at rest in the rotating frame. In that case $\Delta W_{\text{rot}}^{\text{ext}} = \Delta W_{\text{rot}}^{\text{int}} = 0$. Also, $\Delta \sum \frac{1}{2} m_i (\boldsymbol{v}_i)_{\text{rot}}^2 = 0$ and $\Delta \boldsymbol{\omega} \cdot (\boldsymbol{L})_{\text{rot}} = 0$. Consequently, the work–energy theorem for a rigid body is obtained to be

$$\int_1^2 \boldsymbol{N}^{\text{ext}} \cdot d\boldsymbol{\theta} = \Delta(\frac{1}{2}\boldsymbol{\omega} \cdot \overleftrightarrow{\boldsymbol{I}} \cdot \boldsymbol{\omega}).$$

Under the conditions explained in T23, this reduces to

$$\int_1^2 \boldsymbol{N}^{\text{ext}} \cdot d\boldsymbol{\theta} = \Delta((1/2) I \omega^2)$$

which is a result that you must be familiar with.

S32 *The work–energy theorem in a rotating frame that you have derived [Eq. (19)] looks very messy in contrast to its simple form in the laboratory frame. Why should this be so?*

T32 The work–energy theorem in any non-inertial frame, such as a rotating or an accelerating frame, is expected to be different from its form in an inertial frame. This is because, in a non-inertial frame, we have to include the work done by the pseudo-forces acting on a system. (The CM frame, however, happens to be an exception, since as we saw in T23 Ch.7, the work done by the pseudo-forces in this frame adds up to zero.)

S33 *Is it possible to relate the extra terms in Eq. (19) to the work done by the pseudo-forces in the rotating frame ?*

T33 Yes. Let us rewrite Eq. (19) as

$$\Delta T_{\text{rot}} = \Delta W_{\text{rot}}^{\text{ext}} + \Delta W_{\text{rot}}^{\text{int}} + \int \boldsymbol{N}^{\text{ext}} \cdot d\boldsymbol{\theta} - \Delta((1/2)\boldsymbol{\omega} \cdot \overleftrightarrow{\boldsymbol{I}} \cdot \boldsymbol{\omega}) - \Delta(\boldsymbol{\omega} \cdot \boldsymbol{L}_{\text{rot}}). \tag{20}$$

The last three terms add up to the work done by the pseudo-forces. To see this, we must start with the basic equation of motion in a rotating frame

$$m_i \left(\frac{d\boldsymbol{v}_i^{\text{rot}}}{dt}\right)_{\text{rot}} = \boldsymbol{F}_i^{\text{ext}} + \boldsymbol{F}_i^{\text{int}} - m_i \boldsymbol{\omega} \times (\boldsymbol{\omega} \times \boldsymbol{r}_i) - 2m_i \boldsymbol{\omega} \times \boldsymbol{v}_i^{\text{rot}} - m_i \frac{d\boldsymbol{\omega}}{dt} \times \boldsymbol{r}_i.$$

Take dot product with $(d\boldsymbol{r}_i)_{\text{rot}} = \boldsymbol{v}_i^{\text{rot}} dt$ and integrate to get

$$\Delta T_{\text{rot}} = \Delta W_{\text{rot}}^{\text{ext}} + \Delta W_{\text{rot}}^{\text{int}}$$

$$-\int_1^2 \sum_{i=1}^N m_i[\boldsymbol{\omega} \times (\boldsymbol{\omega} \times \boldsymbol{r}_i)] \cdot \boldsymbol{v}_t^{\text{rot}} dt - \int_1^2 \frac{d\boldsymbol{\omega}}{dt} \cdot \boldsymbol{L}_{\text{rot}} dt. \tag{21}$$

Note that the Coriolis force, being always normal to $\boldsymbol{v}_i^{rot}$, does not do any work. We now show that the last three terms in Eq. (20) combine to give the two extra terms on the right side of the above equation.

We have

$$\int_1^2 \boldsymbol{N}^{ext} \cdot d\boldsymbol{\theta} = \int_1^2 \boldsymbol{N} \cdot d\boldsymbol{\theta}$$

$$= \int_1^2 \sum_{i=1}^N \boldsymbol{r}_i \times \left[m_i \left(\frac{d\boldsymbol{v}_i^{\text{rot}}}{dt} \right)_{\text{rot}} + m_i \boldsymbol{\omega} \times (\boldsymbol{\omega} \times \boldsymbol{r}_i) \right.$$

$$\left. + 2m_i \boldsymbol{\omega} \times \boldsymbol{v}_i^{rot} + m_i \frac{d\boldsymbol{\omega}}{dt} \boldsymbol{r}_i \right] \cdot \boldsymbol{\omega} dt \tag{I}$$

and

$$-\Delta \left(\frac{1}{2} \boldsymbol{\omega} \cdot \overleftrightarrow{\boldsymbol{I}} \cdot \boldsymbol{\omega} \right) = -\int \frac{d\boldsymbol{\omega}}{dt} \cdot \overleftrightarrow{\boldsymbol{I}}\, \boldsymbol{\omega} dt - \frac{1}{2} \int \boldsymbol{\omega} \cdot \frac{d\overleftrightarrow{\boldsymbol{I}}}{dt} \cdot \boldsymbol{\omega} dt. \tag{22}$$

Now

$$\frac{d\overleftrightarrow{\boldsymbol{I}}}{dt} = \frac{d}{dt} \sum_{i=1}^N m_i (r_i^2 \boldsymbol{I} - \boldsymbol{r}_i \boldsymbol{r}_i)$$

$$= \sum_{i=1}^N 2m_i [(\boldsymbol{r}_i \cdot \boldsymbol{v}_i) \boldsymbol{I} - \boldsymbol{v}_i \boldsymbol{r}_i].$$

Therefore

$$-\Delta((1/2)\boldsymbol{\omega} \cdot \overleftrightarrow{\boldsymbol{I}} \cdot \boldsymbol{\omega}) = -\int \frac{d\boldsymbol{\omega}}{dt} \cdot \overleftrightarrow{\boldsymbol{I}} \cdot \boldsymbol{\omega} dt - \int \sum_{i=1}^N m_i \boldsymbol{r}_i \cdot \boldsymbol{v}_i^{rot} \omega^2 dt$$

$$+ \int \sum_{i=1}^N m_i (\boldsymbol{\omega} \cdot \boldsymbol{v}_i^{\text{rot}})(\boldsymbol{r}_i \cdot \boldsymbol{\omega}) dt. \tag{II}$$

Note that we have not specified in Eqs. (22) the frame with respect to which the time derivative is to be calculated. This is not necessary since $\boldsymbol{\omega}\cdot \overleftrightarrow{I} \cdot\boldsymbol{\omega}$ is a scalar. You can check explicitly that in Eq. (II) it is immaterial whether we write $\boldsymbol{v}_i$ or $\boldsymbol{v}_i^{\text{rot}}$, since $\boldsymbol{v}_i = \boldsymbol{v}_i^{\text{rot}} + \boldsymbol{\omega}\times\boldsymbol{r}_i$. The same remark is true for the next term to be considered:

$$-\Delta(\boldsymbol{\omega}\cdot\boldsymbol{L}_{\text{rot}}) = -\int \frac{d\boldsymbol{\omega}}{dt}\cdot\boldsymbol{L}^{\text{rot}}dt - \int \boldsymbol{\omega}\cdot\frac{d\boldsymbol{L}^{\text{rot}}}{dt}dt$$

$$= -\int \frac{d\boldsymbol{\omega}}{dt}\cdot\boldsymbol{L}^{\text{rot}}dt - \int \boldsymbol{\omega}\cdot\sum_{i=1}^{N} m_i\boldsymbol{r}_i \times \left(\frac{d\boldsymbol{v}_i^{\text{rot}}}{dt}\right)_{\text{rot}} dt. \quad \text{(III)}$$

It is easy to see that the terms involving $(d\boldsymbol{v}_i^{\text{rot}}/dt)_{\text{rot}}$ in (I) and (III) cancel. Similarly, the $d\boldsymbol{\omega}/dt$ terms in (I) and (II) add up to zero. The second term in (I) trilinear in $\boldsymbol{\omega}$ vanishes identically. The remaining terms then are seen to be equal to the two extra terms on the right side of Eq. (21). We have thus proved that the modification of the usual work–energy theorem that arises in a rotating frame is due to the work done by the centrifugal force and the 'angular acceleration force' in a rotating frame.

9

Illustrative Problems

In this chapter we work out some problems that illustrate and amplify some of the points discussed in this and the earlier volumes. For convenience we depart from the format of teacher-student dialogue adopted so far. Some of the problems discussed here are standard problems that may be familiar to the reader. Our purpose, however, is not just to solve the problems but to point out, wherever possible, the lessons that one learns from them and to warn against possible pitfalls.

Kinematics

P1 *A disc, with polished grooves along OP_1, OP_2, ..., OP_N, is fixed in a vertical plane as shown in Fig. 9.1. N ball-bearings of different masses are released from O simultaneously, one along each groove. Show that all of them reach the edge of the disc simultaneously.*

A1 Suppose the ith chord OP_i makes an angle θ_i with the vertical. The component of acceleration due to gravity along OP_i is $g\cos\theta_i$. Therefore, the time taken by a ball-bearing to travel OP_i is

$$t_i = \sqrt{\frac{2OP_i}{g\cos\theta_i}} = \sqrt{\frac{2D}{g}}$$

(where D is the diameter of the disc), which is independent of i. Thus all the ball-bearings take the same time to reach the edge of the disc.

What is wrong with the above procedure? The vertical distance covered by the ith mass is $OP_i\cos\theta_i$. g is acceleration in the same direction.

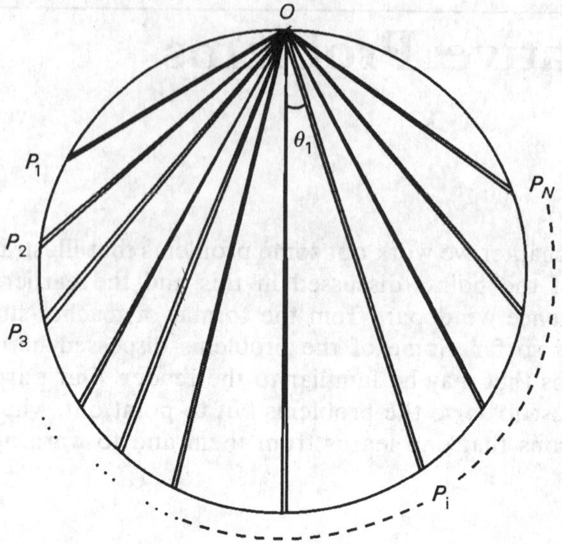

Fig. 9.1

Therefore,

$$t_i = \sqrt{\frac{2OP_i \cos\theta_i}{g}}$$

which depends on i. This is the sort of thing we do in a problem of projectile motion: consider the horizontal and vertical motion separately. Why not here?

The answer is that in this problem motion is subject to <u>two</u> forces, gravity and the groove's reaction. Consequently the acceleration of the ball-bearing in the vertical direction is not g, neither is the horizontal acceleration zero. For a general path OP_i, it is inconvenient to decompose the forces along the vertical and horizontal directions because the groove's reaction needs to be taken into account. It is much wiser to decompose the forces along OP_i and a direction perpendicular to it. As per the constraint of the problem, the component of the force in the latter direction must be zero and the net force along OP_i is only due to gravity.

Of course, if one insists on decomposition along the vertical and horizontal directions, we may proceed thus: the groove's reaction is easily determined to be $mg\sin\theta$. The acceleration components, therefore, are

$$a_h = g\sin\theta_i\cos\theta_i,$$

$$a_v = g - g\sin^2\theta_i = g\cos^2\theta_i.$$

The time taken to travel a vertical distance $OP_i\cos\theta_i$ is then given by

$$t_i = \sqrt{\frac{2OP_i\cos\theta_i}{g\cos^2\theta_i}} = \sqrt{\frac{2D}{g}},$$

as before.

Note that the mass of the ball-bearing enters nowhere in the problem. This interesting result was first emphasized by Galileo. In modern terms it shows the equivalence of inertial and gravitational mass.

P2 *Figure 9.2 shows two places P and Q situated on the two sides of a river of width ω. A man who can run r km/h and swim in still water at s km/h wishes to go from P to Q in the shortest possible time. Determine the path he must follow and the time taken to go along that path. The speed of the river flow is equal to the swimming speed of the man, viz. s km/h, from right to left. The other bank of the rivers is not walkable.*

A2 Let Q' be the point directly in front of Q across the river and let d be the distance PQ'. The man can first run a distance x along the bank up to a point M, and then swim straight across to Q. Clearly, while swimming, the net velocity of the man relative to the bank ($\boldsymbol{v}_{mb}$) must be along MQ. Let the velocity of the man relative to the river water ($\boldsymbol{v}_{mr}$) make an angle θ with the river flow. If t is the time (in hours) required for swimming,

$$(|\boldsymbol{v}_{mr}|\cos\theta + s)t = d - x,$$

$$|\boldsymbol{v}_{mr}|\sin\theta t = \omega, \tag{A2.1}$$

which gives, since $|\boldsymbol{v}_{mr}| = s$,

$$\tan\frac{\theta}{2} = \frac{\omega}{d-x}.$$

Clearly, $-\infty < x < d$ since $0 < \theta < \pi$; the swimmer cannot reach Q from a point ahead of Q'. This conclusion is not generally true; it is a consequence of the fact that, in the present problem, the swimming speed of the man and the flow speed are equal.

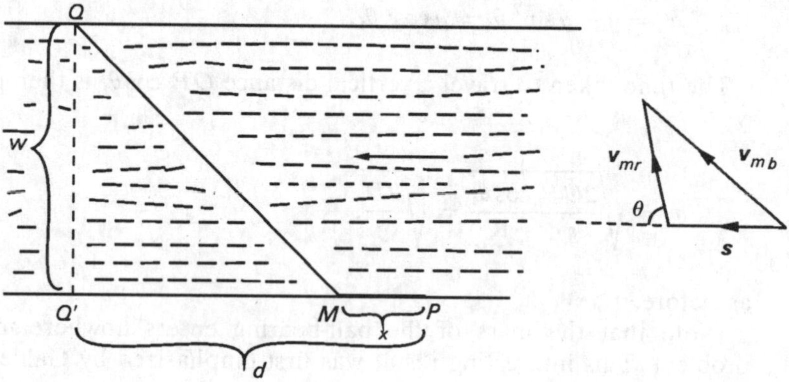

Fig. 9.2

Equation (A2.1) also gives

$$t = \frac{\omega^2 + (d-x)^2}{2s(d-x)}.$$

The total time for the path τ is given by

$$\tau = \frac{x}{r} + \frac{\omega^2 + (d-x)^2}{2s(d-x)}.$$

In order to determine the minimum time $\tau_{\min}$, we calculate $d\tau/dt$,

$$\frac{d\tau}{dt} = \frac{1}{r} - \frac{1}{2s} + \frac{\omega^2}{2s(d-x)^2}.$$

Clearly, for $r \le 2s$, $d\tau/dx > 0$ and τ is monotonic increasing. When $r > 2s$, τ has a minimum at $x = d - w\sqrt{[r/(r-2s)]}$. (Note that x can be positive or negative.) Hence,

$$\tau_{\min} = \frac{d}{r} + \frac{\omega}{s}\sqrt{\frac{r-2s}{r}}.$$

The point of the problem is that naive intuitive expectations can sometimes be misleading. The fact that one may have to walk away from the destination ($x < 0$) in order to reach it in the least time may come as a surprise. But a little reflection should convince us that, when the flow speed is high, it is necessary to start from a point further upstream if one is not to be swept by the flow past the point Q!

Static Equilibrium

P3 *A semicircular rod of mass m and radius r is hinged at A by a frictionless pin. Show that, in equilibrium, the angle made by the diameter AB [shown in Fig. 9.3(a)] with the vertical is independent of r. What force in the horizontal direction should be applied at B to have the diameter AB vertical? Determine the reaction of the hinge in the latter case.*

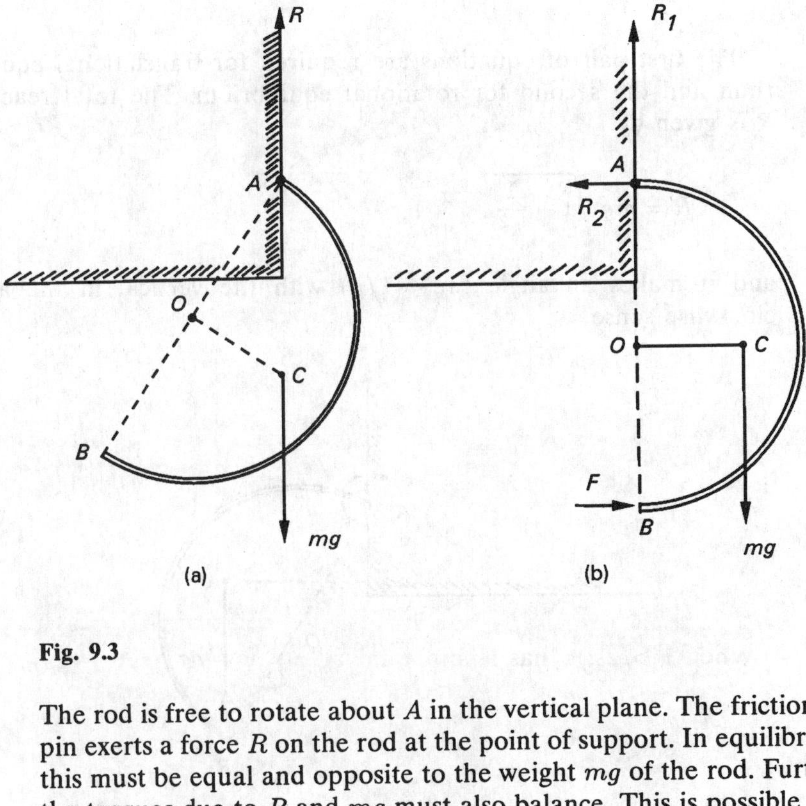

Fig. 9.3

A3 The rod is free to rotate about A in the vertical plane. The frictionless pin exerts a force R on the rod at the point of support. In equilibrium, this must be equal and opposite to the weight mg of the rod. Further, the torques due to R and mg must also balance. This is possible only if the two forces have a common line of action. Thus, in equilibrium, the CM of the rod C must be vertically below A.

Now, the distance of the CM of a uniform semicircular rod from its centre O can be shown to be equal to $2r/\pi$. Therefore the angle θ which the diameter AB makes with the vertical is given by

$$\tan\theta = \frac{2r/\pi}{r} = \frac{2}{\pi}.$$

Let F be the force applied at B so that the diameter AB is vertical

[Fig. 9.3(b)]. R_1, and R_2 are the vertical and horizontal components of the reaction at A. Equilibrium of the rod implies

$$mg = R_1, \quad F = R_2,$$

and

$$mg \times \frac{2r}{\pi} = F \times 2r,$$

i.e. $$F = \frac{mg}{\pi}.$$

The first pair of equations are required for translational equilibrium and the second for rotational equilibrium. The total reaction $\boldsymbol{R}$ is given by

$$R = mg\sqrt{1 + \frac{1}{\pi^2}}$$

and it makes an angle $\tan^{-1}(1/\pi)$ with the vertical in the anticlockwise sense.

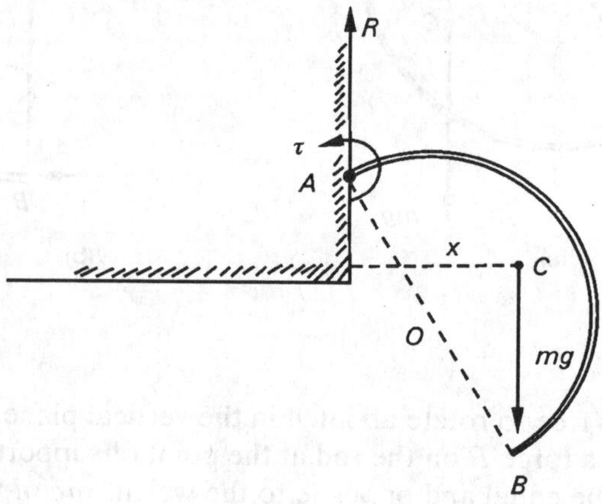

Fig. 9.4

What do we learn from this problem? The frictionless hinge constrains the end of the rod from translational motion; it is, however, free to rotate about the point support. The reaction of a frictionless hinge is a force $\boldsymbol{R}$ acting at the point of support. <u>*The magnitude and direction of* **R** *are not a priori fixed*</u>; R is a self-adjusting force determined by the equilibrium conditions. The rod can be fully constrained, from translational as well as rotational motion, by fixing it rigidly to the support as shown in Fig. 9.4.

Obviously, it can be fixed in any orientation. How does one explain this equilibrium? The answer is that the support, in this case, not only provides a self-adjusting reaction $\boldsymbol{R}$, but it also provides a counter-torque τ. In Fig. 9.4, $R = mg$ and τ has a magnitude equal to mgx and acts in the anti-clockwise sense. You should appreciate that the fixed support at A is not quite a point, so it is capable of giving rise to a configuration of contact forces that has non-zero torque about A. In short, the nature of support determines the type of reaction it can provide.

P4 *The lid of a box is hinged at one end. The box is kept open by supporting the lid on a stick hinged at the opposite end as shown in Fig. 9.5(a). Show that equilibrium is possible only if the coefficient of static friction between the stick and the lid, $\mu \geq \tan\theta/2$, where θ is the angle of the lid with the horizontal. Assume the stick to be massless.*

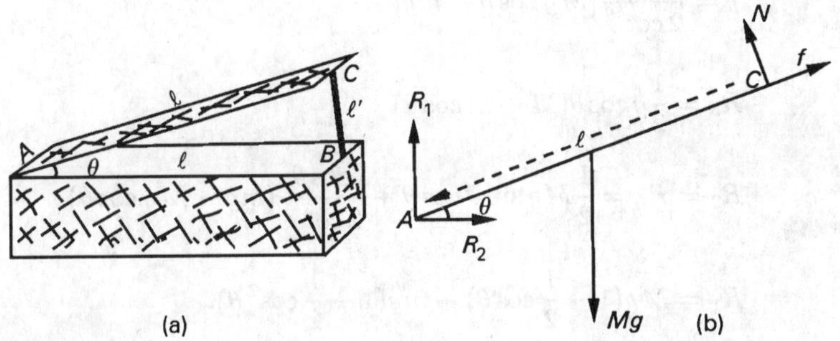

Fig. 9.5

A4 Consider the equilibrium of the lid of mass M and length ℓ. The free-body diagram is shown in Fig. 9.5(b). N and f are the components of the force exerted by the stick on the lid at C. R_1 and R_2 are the components of the force exerted by the hinge on the lid at A. The equilibrium conditions are

$$\left.\begin{aligned} R_1 + N\cos\theta + f\sin\theta &= Mg \\ R_2 + f\cos\theta &= N\sin\theta \end{aligned}\right\} \quad \textit{translational equilibrium} \qquad \text{(A4.1)}$$

$$N \times \ell = Mg \times \frac{\ell}{2}\cos\theta \quad \textit{rotational equilibrium}$$

The three equations involve four unknowns: R_1, R_2, N and f. Now consider the equilibrium of the stick of mass m and length ℓ'. (We shall later take $m \to 0$.) The free-body diagram of the stick is shown in Fig. 9.6. The force on the stick by the lid at C is equal and opposite to that by the stick on the lid. The components of the force on the stick by the hinge at B, R_3 and R_4, are additional unknowns. The equilibrium of the stick implies

$$\left.\begin{aligned} mg + N\cos\theta + f\sin\theta = R_3 \\ R_4 + f\cos\theta = N\sin\theta \end{aligned}\right\}\ \textit{translational equilibrium} \qquad \text{(A4.2)}$$

$$\left. mg \times \frac{\ell'}{2}\sin\frac{\theta}{2} + f \times \ell'\cos\frac{\theta}{2} = N \times \ell'\sin\frac{\theta}{2} \right\}\ \textit{rotational equilibrium}$$

(Note that sides AB and AC are equal, so the angles made by N and R_3 with BC are $\theta/2$.)

We now have six equations involving six unknowns: N, f, R_1, R_2, R_3 and R_4. These can be solved to obtain

$$N = \frac{1}{2}Mg\cos\theta$$

$$f = \frac{1}{2\cot\frac{\theta}{2}}[Mg\cos\theta - mg]$$

$$R_3 = \frac{1}{2}g\cos\theta[M + m\cos\theta)$$

$$R_2 = R_4 = \frac{1}{2}Mg\sin\theta\cos\theta + \frac{\cos\theta}{2\cot\frac{\theta}{2}}(mg - Mg\cos\theta)$$

$$R_1 = Mg(1 - \frac{1}{2}\cos\theta) + mg(1 - \frac{1}{2}\cos^2\theta).$$

For $m = 0, f/N = \tan\theta/2$. The limiting friction $f_s = \mu_s N$. Since $f \le f_s$, $\tan\theta/2 \le \mu_s$.

What if we consider the lid + stick system as a whole? The forces at C are internal forces of the system. They cancel in pairs and do not appear in the equilibrium conditions of the total system:

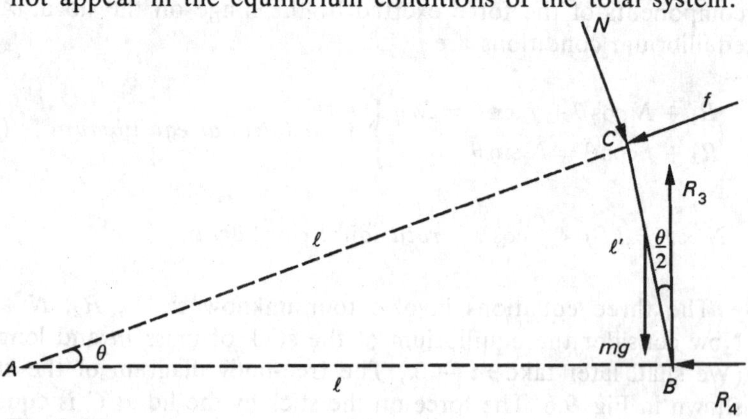

Fig. 9.6

$$R_1 + R_3 = (M+m)g$$
$$R_2 = R_4$$
$$R_3 \times \ell = Mg \times \frac{\ell}{2}\cos\theta + mg\left[\ell - \frac{\ell'}{2}\sin\frac{\theta}{2}\right]$$

which are contained in Eqs. (A4.1) and (A4.2). Note that the total structure ACB is not a single rigid body but a system of two connected rigid bodies; the angle between the lid and the stick can vary. Hence, the three equations above are not enough to characterize its equilibrium.

If a massless rod, with two forces acting at its two ends, is in equilibrium, then the two forces must be equal and opposite and they must act along the length of the rod. This may be verified in the present problem by seeing that the resultant of N and f has the same magnitude as that of R_3 and R_4. Also, the resultant of N and f acts along the rod since $f/N = \tan\theta/2$. Obviously the resultant of R_3 and R_4 must also act along the rod in the opposite direction; it can be easily verified that $R_4/R_3 = \tan\theta/2$.

P5 *A thin non-uniform rod of length L and mass M has its CM at a distance αL from one of its ends. It rests inside a smooth hemispherical bowl of radius R as shown in Fig. 9.7(a). Determine the angle made by the rod with the horizontal.*

A5 This example illustrates a simple rule about the equilibrium of a rigid body. If a rigid body is in equilibrium under the action of three forces acting at three different points, the lines of action of the forces are either parallel or pass through a common point (i.e. they are concurrent).

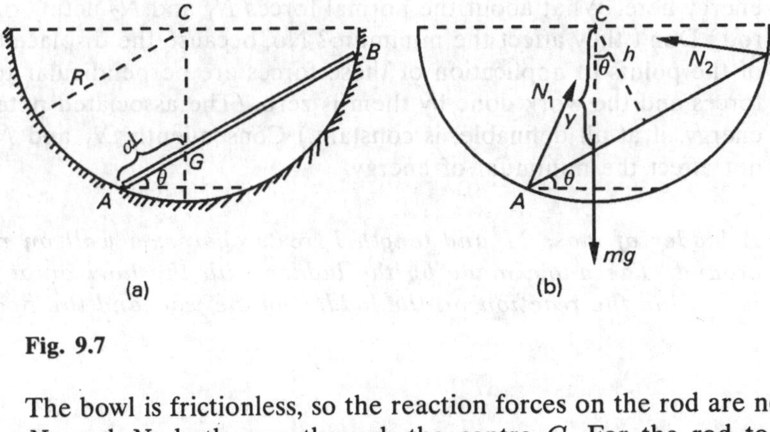

Fig. 9.7

The bowl is frictionless, so the reaction forces on the rod are normal; N_1 and N_2 both pass through the centre C. For the rod to be in equilibrium, therefore, the vertical line of action of the weight of the rod must also pass through C. In other words, the CM of the rod must be vertically below C. From the geometry of Fig. 9.7(b),

$$h = \sqrt{R^2 - \frac{L^2}{4}}$$

and

$$\tan\theta = \frac{\frac{L}{2} - \alpha L}{\sqrt{R^2 - \frac{L^2}{4}}}.$$

Note that the angle θ does not depend on the mass of the rod; it depends only on the position of the CM along the rod. The magnitudes of N_1 and N_2 depend on the mass of the rod. This problem can also be solved by minimizing the potential energy of the rod as follows: Let us take the reference level for the calculation of potential energy at the bottom of the bowl [see Fig. 9.7(b)]. The angle made by the line joining the centre of the rod and the centre of the bowl with the vertical is θ. The potential energy of the rod $V(\theta)$ is

$$\begin{aligned} V(\theta) &= Mg(R - y) \\ &= MgR - Mg\left[h\cos\theta + \left(\frac{1}{2} - \alpha\right) L \sin\theta\right]. \end{aligned}$$

(Note that now θ is a variable.) The minimization requirement $dV/d\theta = 0$ gives $\tan\theta = (1/2 - \alpha)L/h$, which is the same as obtained before.

Note that we have considered only the gravitational potential energy here. What about the normal forces N_1 and N_2 acting on the rod? Don't they affect the minimum? No, because the displacements of the points of application of these forces are perpendicular to the forces and the work done by them is zero. (The associated potential energy, if at all definable, is constant.) Consequently N_1 and N_2 do not affect the minimum of energy.

P6 *A ladder of mass M and length L rests against a wall on rough ground. The angle made by the ladder with the horizontal floor is θ. Find the reaction on the ladder by the wall and the floor.*

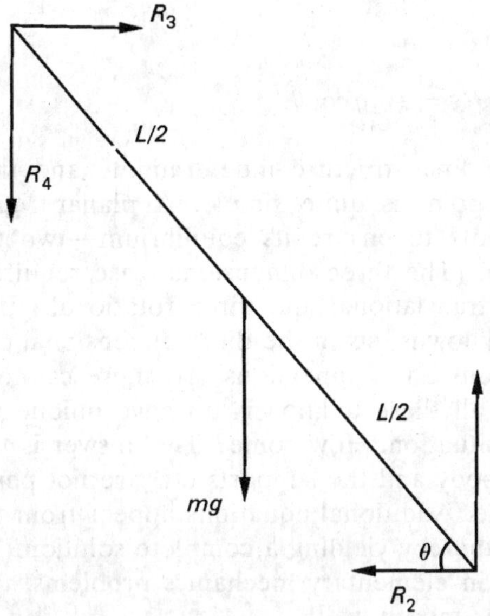

Fig. 9.8

A6 It is not stated in the problem whether the wall is smooth or rough. Let us first consider the wall to be rough. Figure 9.8 shows the free-body diagram of the ladder. R_1, R_2 are the components of the force on the ladder by the ground and R_3 and R_4 are those of the force by the wall. Equilibrium conditions imply

$$\left.\begin{aligned} R_3 + R_2 &= 0 \\ R_1 &= mg + R_4 \end{aligned}\right\} \textit{translational equilibrium}$$

$$mg \times \frac{L}{2} \cos\theta + R_2 \times L \sin\theta = R_1 \times L \cos\theta \qquad \textit{rotational equilibrium}$$

These are three equations in four unknowns: R_1, R_2, R_3 and R_4. Clearly all the unknowns cannot be determined.

Let us now suppose that the wall is smooth so that $R_4 = 0$. The equilibrium conditions now give

$$\left.\begin{aligned} R_3 + R_2 &= 0 \\ R_1 &= mg \end{aligned}\right\} \textit{translational equilibrium}$$

$$mg \times \frac{L}{2} \cos\theta + R_2 \times L \sin\theta = R_1 \times L \cos\theta \ \textit{rotational equilibrium}$$

These are three equations in three unknowns and can be solved easily to obtain

$$R_1 = mg$$

$$R_2 = -R_3 = \tfrac{1}{2} mg \cot\theta.$$

Why is the first structure indeterminate and the second determinate? The point is quite simple. A planar rigid body requires three constraints to ensure its equilibrium—two translational and one rotational. (The three dimensional case requires six constraints in all—three translational and three rotational.) If there are more than three unknowns (six in the three-dimensional case) arising from various reactions and connections etc. these cannot be determined uniquely. Yet all these unknowns do have unique values in the actual physical situation. How come? The answer is that in the actual situation the body and the supports etc. are not perfectly rigid; they are deformable. Additional equations appear from the details of the deformation, thereby yielding a complete solution. You can now appreciate why, in elementary mechanics problems, the ladder always rests against a smooth wall!

Stability of Equilibrium

P7 *A solid hemisphere of radius R is attached to a solid cylinder of radius R and height h. The composite body has a uniform mass density throughout. It is kept on a horizontal surface as shown in Fig. 9.9(a). What is the maximum value of the ratio h/R for which the body is in stable equilibrium? Assume that the body rolls without slipping.*

A7 If a horizontal force is applied to the cylinder, it will slide if there were no friction. For the cylinder to roll without slipping there must be sufficient friction between the bottom of the cylinder and the surface. Suppose that an external force rolls the cylinder slightly, so that its angular displacement is θ, and it is in equilibrium in the displaced position as shown in Fig. 9.9(a). Clearly, the net torque on the cylinder in this position is zero. Suppose that the external force is now withdrawn so that the cylinder experiences an unbalanced torque. If, for small θ, this torque tends to bring the cylinder back to the original position, the equilibrium is stable.

The forces on the cylinder in Fig. 9.9(b) are: The reaction N at the point of contact M and acting along a line perpendicular to the surface (passing through the centre O of the hemisphere), the frictional force f at M acting along the horizontal surface, and the weight Mg acting vertically at the centre of mass C. Taking moments about M it is easily seen that the net torque restores the cylinder to the original position if the line of action of weight is to the left of M. Thus the equilibrium is stable (unstable) if AC is less than (greater than) $AO = R$. The problem, therefore, reduces to locating the CM of the body.

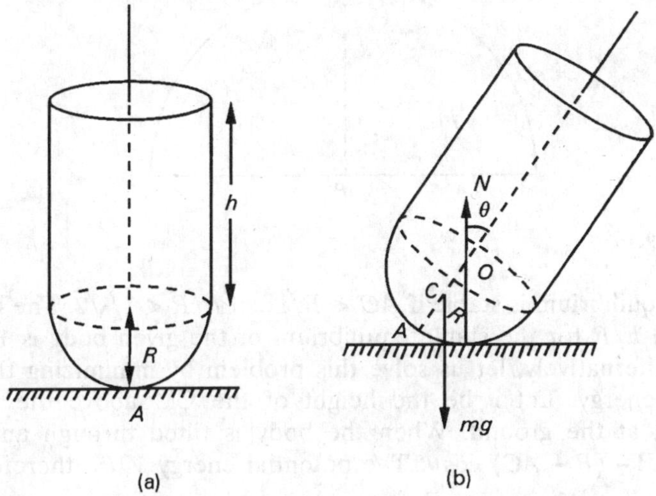

Fig. 9.9

First, let us determine the CM of a hemisphere of density ρ and radius R. By symmetry it must lie on the line PQ (see Fig. 9.10), which we take to be the z-axis of spherical polar co-ordinates.

The CM is given by

$$z_{CM} = \frac{1}{M}\int z\rho dv$$

$$= \frac{1}{\left(\frac{2}{3}\pi R^3\right)}\int_0^R\int_0^{\pi/2}\int_0^{2\pi} r\cos\theta r^2 dr\, d\theta\, d\phi = \frac{3R}{8}\,.$$

i.e. the CM of the hemisphere lies at a distance $5R/8$ away from Q on PQ.
The CM of the body in Fig. 9.9(b) can now be found:

$$AC = \frac{\frac{5R}{8}\rho\frac{2\pi}{3}R^3 + \left(R + \frac{h}{2}\right)\rho\pi R^2 h}{\rho\left(\frac{2\pi}{3}R^3 + \pi R^2 h\right)}$$

$$= \frac{\frac{5}{12}R^2 + \left(R + \frac{h}{2}\right)h}{\frac{2R}{3} + h}$$

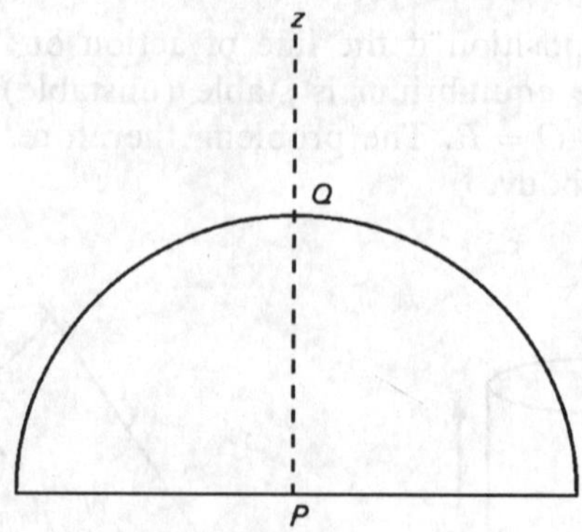

Fig. 9.10

Equilibrium is stable if $AC < R$, i.e. if $h/R < 1/\sqrt{2}$. The maximum ratio h/R for the stable equilibrium of the given body is $1/\sqrt{2}$.

Alternatively, let us solve this problem by minimizing the potential energy. Let z be the height of the CM above the reference level at the ground. When the body is tilted through an angle θ, $z = R - (R - AC)\cos\theta$. The potential energy $V(\theta)$, therefore, is

$$V(\theta) = Mg[R - (R - AC)\cos\theta].$$

Clearly,

$$\frac{dV}{d\theta} = (R - AC)\sin\theta = 0 \quad \text{at} \quad \theta = 0$$

$$\frac{d^2V}{d\theta^2} = (R - AC)\cos\theta = R - AC \quad \text{at} \quad \theta = 0.$$

The potential energy is thus minimum at $\theta = 0$ for $AC < R$ and maximum for $AC > R$ corresponding to stable and unstable equilibrium, respectively.

Tilting

P8 *A rectangular box of length ℓ breadth b and height h rests on a plane surface. The plane is hinged at one end so that it can be tilted with respect to the horizontal. Determine the condition that the box topples over before beginning to slide down the plane, as the inclination of the plane is increased. The coefficient of static friction between the two surfaces is μ.*

A8 The translational equilibrium of the box on the plane, when the plane makes an angle θ with the horizontal (see Fig. 9.11), requires

$$f = mg\sin\theta$$

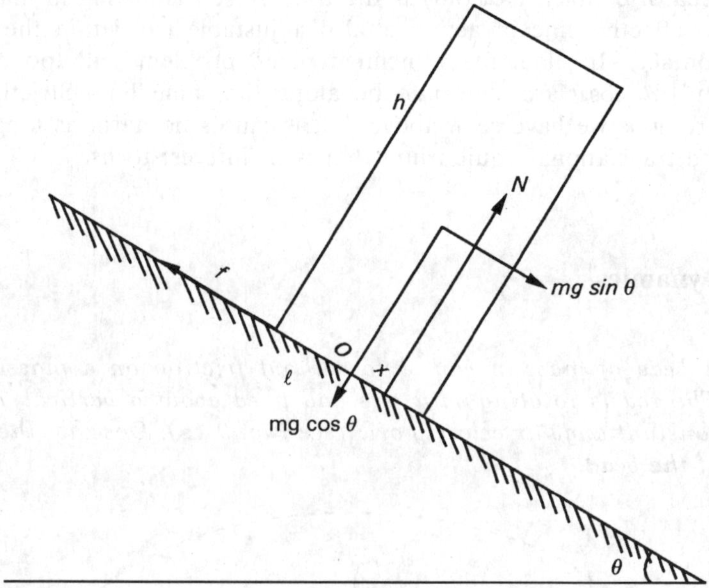

Fig. 9.11

where f is the frictional force, and

$$N = mg\cos\theta.$$

Now, $\quad f_{\max} = \mu N = \mu mg\cos\theta.$

Therefore, translational equilibrium is possible up to $\theta = \theta_t$, where $\tan\theta_t = \mu$.

To investigate rotational equilibrium, let us calculate the torques about O. Clearly, the torque of f is zero. The torque of the reaction N must, therefore, be equal and opposite to that of the weight. So, supposing that N acts at a distance x from O, we must have

$$mg \times \frac{h}{2}\sin\theta = mg\cos\theta \times x,$$

i.e. $\quad x = \dfrac{h}{2}\tan\theta$

Since $x_{\max} = \ell/2$, rotational equilibrium is possible (subject to the existence of translational equilibrium) up to $\theta = \theta_r$ where $\tan\theta_r = \ell/h$.

If $\theta_r < \theta_t$, i.e. if $\ell/h < \mu$, the box tilts before it begins sliding down. Otherwise it will slide without toppling.

This problem is meant to point out an obvious but little noticed fact: the normal reaction N is a distributed force acting over the area of contact. Not only is the total N self-adjusting in magnitude, its effective line of action is also adjustable (to within the area of contact). In elementary inclined plane problems, all too often, N and $mg\cos\theta$ are shown to be along the same line. Strictly this is wrong as we have seen above, but it causes no harm as long as it is the translational equilibrium that is of interest to us.

Dynamics

P9 *A bead of mass m can slide without friction on a massless rod. The rod is rotating with one end fixed about a vertical axis at a constant angular velocity of ω (see Fig. 9.12). Describe the motion of the bead.*

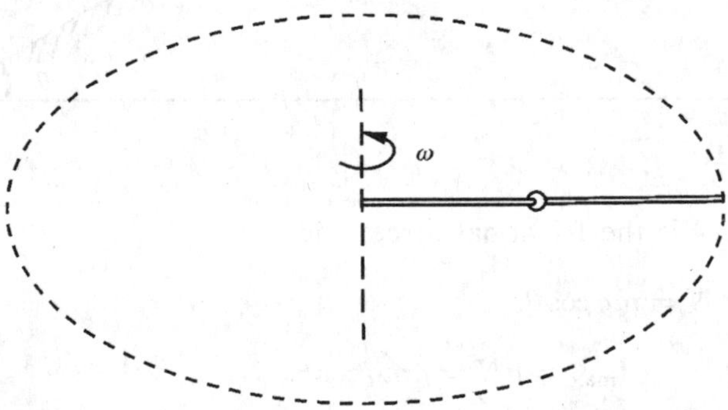

Fig. 9.12

A9 The motion is confined to a horizontal plane; the weight of the bead and the vertical component of the reaction by the rod on the bead balance each other. The horizontal component $\boldsymbol{F}$ of the force by the rod is the net force on the bead. Let us employ the plane polar co-ordinates r and θ to describe the motion. The position, velocity and acceleration vectors are

$$\boldsymbol{r} = r\widehat{r}$$

$$\boldsymbol{v} = \dot{r}\widehat{r} + r\dot{\theta}\widehat{\theta}$$

$$\boldsymbol{a} = (\ddot{r} - r\dot{\theta}^2)\widehat{r} + (2\dot{r}\dot{\theta} + r\ddot{\theta})\widehat{\theta}.$$

Since the bead slides without friction, the force $\boldsymbol{F}$ is normal to the length of the rod in the plane of motion:

$$\boldsymbol{F} = F\widehat{\theta}.$$

The equations of motion for the bead, therefore, are

$$0 = m(\ddot{r} - r\omega^2) \tag{A9.1}$$

$$F = m(2\dot{r}\omega).$$

In order to solve the first equation we multiply it by $\dot{r}dt$ and rearrange it to obtain

$$d\left(\frac{1}{2}m\dot{r}^2 - \frac{1}{2}mr^2\omega^2\right) = 0$$

Thus

$$\dot{r}^2 - r^2\omega^2 = \textit{constant}.$$

The constant can be determined assuming the initial conditions $r = r_0$ and $\dot{r} = 0$ at $t = 0$. We thus obtain

$$\dot{r} = \omega\sqrt{r^2 - r_0^2}.$$

This equation can be integrated assuming the same initial conditions. The result is

$$\log\frac{r + \sqrt{r^2 - r_0^2}}{r_0} = \omega t,$$

$$\text{i.e. } r = \frac{r_0}{2}\left(e^{\omega t} + e^{-\omega t}\right).$$

Having solved for $\dot{r}$, the second part of Eq. (A9.1) gives the horizontal force by the rod on the bead

$$\boldsymbol{F} = 2m\omega^2\sqrt{r^2 - r_0^2}\widehat{\theta}.$$

Why do we employ the polar co-ordinates instead of the familiar cartesian co-ordinates for solving this problem? In terms of the latter the equations of motion apparently have a simple form: $F_x = md^2x/dt^2$, $F_y = md^2y/dt^2$. The difficulty, however, is that both F_x and F_y are unknown and non-zero. In polar co-ordinates, in contrast, one component of the reaction is identically zero because the reaction is always normal to the rod. Consequently, one of the equations (the radial equation) has no unknowns and can be readily solved.

It is fun getting the same equations using the rotating frame of the rod. In this non-inertial frame one must include the centrifugal force and the Coriolis force. In the rotating frame the rod is in a fixed direction say x'. The horizontal reaction of the rod is then along y', i.e. $\boldsymbol{F} = F\hat{j}'$. The angular velocity is in the vertical (z or z') direction. The centrifugal force is $m\omega^2 x'\hat{i}'$ and the Coriolis force is $-2m\omega\dot{x}'\hat{j}'$. The equations of motion for the bead, in this frame, therefore, are

$$m\frac{d^2x'}{dt^2} = m\omega^2 x'$$

$$0 = F - 2m\omega\dot{x}'.$$

Remembering that x' in this frame is r in the laboratory frame, these equations are identical to Eq. (A9.1).

Variable Mass Systems

P10 *A chain of length L is held vertically with its lower end touching the top of a table. When released, it gathers on the table in a heap of negligible size. Calculate the time in which the entire chain drops on the table?*

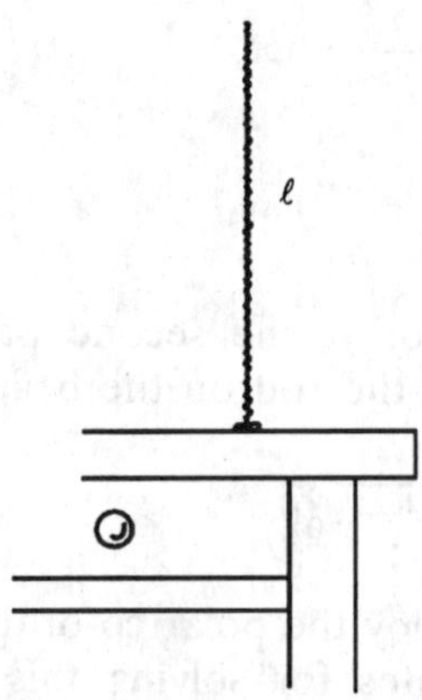

Fig. 9.13

A10 We saw in Ch.4, Box. 4.2 that, for a variable mass system, the equation of motion is

$$m\frac{d\boldsymbol{v}}{dt} = \boldsymbol{F}_m^{ext} + \boldsymbol{F}_{m,\delta m}, \tag{A10.1}$$

where

$$\boldsymbol{F}_{m,\delta m} = \boldsymbol{F}_{\delta m}^{ext} - |\frac{dm}{dt}|(\boldsymbol{v}_a - \boldsymbol{v}_b). \tag{A10.2}$$

In these equations m is the instantaneous mass of the system and δm is the mass element entering or leaving the system. $\boldsymbol{v}_a$ and $\boldsymbol{v}_b$ are the velocities of δm after and before entering/leaving the system respectively.

In the present problem, the mass element δm at the lower end of the falling chain comes to rest instantaneously upon hitting the table. Thus $\boldsymbol{F}_{\delta m}^{\text{ext}}$ is the impulsive reaction force by the table. The chain, nonetheless, is falling freely under gravity; the tension throughout the chain is zero. Consequently, $\boldsymbol{F}_{m,\delta m} = 0$ in the equations above and the two equations decouple. One describes the motion of the chain and the other the motion of the element δm.

Let λ be the mass per unit length of the chain and let ℓ be the length of the vertical part of the chain. Clearly, the instantaneous mass of the system $m = \lambda\ell$. The height of the CM of the chain above the ground, $z = \ell/2$, so that its velocity is $dz/dt = 1/2d\ell/dt$. Equation (A10.1) gives

$$\frac{\lambda\ell}{2}\frac{d^2\ell}{dt^2} = -\lambda\ell g,$$

which can be easily integrated to obtain

$$\frac{d\ell}{dt} = -2gt$$

$$\ell = L - gt^2.$$

where we have assumed the initial conditions: $\ell = L$ and $d\ell/dt = 0$ at $t = 0$. Thus the time of fall for the chain is

$$T = \sqrt{\frac{L}{g}}.$$

Equation (A10.2) may now be used to calculate $F_{\delta m}^{\text{ext}}$. We have

$$\frac{dm}{dt} = \lambda\frac{d\ell}{dt} = -2g\lambda t,$$

$v_a = 0$ and $v_b = v = d\ell/dt = -2gt$. Hence,

$$F_{\delta m}^{\text{ext}} = 4\lambda g^2 t^2.$$

Note, that the total reaction force N of the table on the chain is $N = F_{\delta m}^{ext} + \lambda(L - \ell)g$, where the second term is the usual reaction force on the part already on the table. At the end when $t = \sqrt{(L/g)}$, $F_{\delta m}^{\text{ext}}$ discontinuously changes from $4\lambda Lg$ to zero while the second term becomes λLg. The discontinuity in $F_{\delta m}^{\text{ext}}$ is to be expected because a finite force is needed to bring to rest every infinitesimal element till the end.

This problem is qualitatively different from the familiar variable mass problems about rockets, conveyor belts, etc. There the external force on δm is negligible while $\boldsymbol{F}_{\delta m,m}$ is significant. As we have seen the situation is just the opposite in the present problem. The self-adjusting nature of the reaction force N is also nicely illustrated here.

P11 *A rain drop falls from a height h and increases its size (from zero initial size) by accreting all the mist it encounters. If the mist is of uniform density ρ and stretches up to the ground, determine the velocity and displacement of the rain drop as a function of time?*

A11 For a rain drop gathering mist, δm is the mass of the mist it encounters in time δt. The external force of gravity on δm is clearly negligible. The mist sticking to the falling rain drop changes its velocity instantly from zero to $\boldsymbol{v}$, the instantaneous velocity of the rain drop. So $v_b = 0$ and $\boldsymbol{v}_a = \boldsymbol{v}$. So Eqs. (A10.1) and (A10.2), for this one-dimensional problem, reduce to

$$\frac{d}{dt}(mv) = mg. \qquad \text{(A11.1)}$$

The instantaneous mass of the rain drop is

$$m(t) = \frac{4\pi}{3}r^3(t)\rho_\omega, \qquad \text{(A11.2)}$$

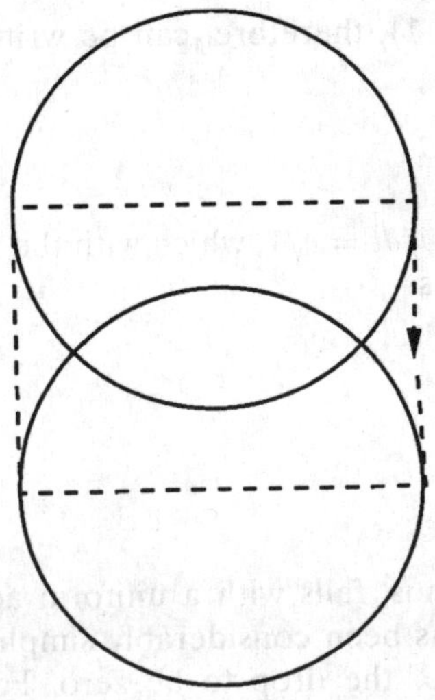

Fig. 9.14

where $r(t)$ is the radius of the rain drop at that instant and ρ_w is the density of water. From Fig. 9.14

$$\delta m(t) = \rho\pi r^2(t)v(t)dt.$$

Therefore $$\frac{dm}{dt} = \rho\pi r^2 \frac{dz}{dt} \qquad \text{(A11.3)}$$

where $v = dz/dt$, z being the vertical co-ordinate of the centre of the drop. Equation (A11.2), on the other hand, gives

$$\frac{dm}{dt} = \rho_w 4\pi r^2 \frac{dr}{dt}.$$

Comparing the two expressions for dm/dt we obtain

$$\frac{dr}{dt} = \frac{\rho}{4\rho_w}\frac{dz}{dt},$$

which can be easily integrated to give

$$r = \frac{\rho}{4\rho_w} z, \qquad \text{(A11.4)}$$

where the initial size of the rain drop at the starting location $z = 0$ has been taken to be zero.

Now, using Eqs. (A11.2), (A11.3) and (A11.4), we obtain

$$\frac{1}{m}\frac{dm}{dt} = \frac{3v}{z}.$$

Equation (A11. 1), therefore, can be written as

$$\frac{3v^2}{z} + \frac{dv}{dt} = g. \tag{A11.5}$$

By inspection, $dv/dt = g/7$, which with the initial conditions $z = 0$, $v = 0$ at $t = 0$ gives

$$v = \frac{g}{7}t$$

$$z = \frac{g}{14}t^2.$$

The raindrop, thus, falls with a uniform acceleration of $g/7$.

This problem has been considerably simplified since we assumed the initial size r_0 of the drop to be zero. For $r_0 \neq 0$, Eq. (A11.5) modifies to

$$\frac{3v^2}{\left(z + \frac{4\rho_w}{\rho} r_0\right)} + \frac{dv}{dt} = g.$$

The solution in this case is far more complicated and is not pursued here.

There are interesting and simple variants of this problem. Suppose the mass accretion rate of the rain drop is constant. The equation of motion in this case is

$$\frac{dv}{dt} + \frac{v}{t} = g.$$

whose solution is $v = (g/2)t$, i.e. the drop falls with a uniform acceleration of $g/2$. If, on the other hand, we take the mass accretion rate to be proportional to the surface area, the equation of motion is found to be

$$\frac{3v}{t} + \frac{dv}{dt} = g,$$

whose solution is $v = (g/4)t$, i.e. the acceleration of the drop is $g/4$.

The moral of the problem is self-evident. The acceleration of a falling raindrop depends on the mechanism of mass accretion that is operative.

Non-inertial frames

P12 *On an inclined plane two objects of masses m_1 and m_2 move down with initial velocities v_{1i} and v_{2i} (along the length of the plane) and collide while on the plane. Determine their velocities after the collision, assumed elastic.*

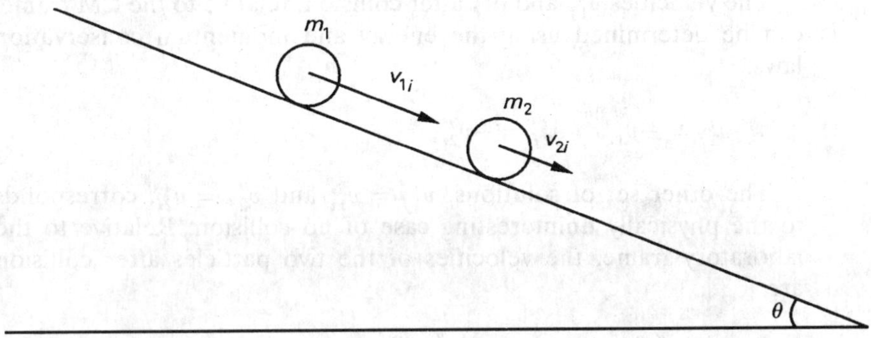

Fig. 9.15

A12 This is a one-dimensional problem; choose the positive x-axis down the inclined plane. Both the masses move down the inclined plane with an acceleration $g \sin\theta$. Their velocities before collision, therefore, are given by

$$v_1 = v_{1i} + g \sin\theta t, \quad v_2 = v_{2i} + g \sin\theta t.$$

The CM of the two masses has the initial velocity

$$V_i^{CM} = \frac{m_1 v_{1i} + m_2 v_{2i}}{m_1 + m_2},$$

and it moves down the plane with velocity

$$V^{CM} = V_i^{CM} + g \sin\theta t.$$

Let us look at the collision relative to the non-inertial CM frame moving with an acceleration $\alpha = g \sin\theta$. The II law of motion in this frame is

$$ma = F - m\alpha.$$

Clearly, the acceleration of both the particles in the CM frame is zero. Therefore, relative to the CM frame, the velocities of the two particles before collision are the same as their initial velocities.

These can be obtained, using the usual velocity addition rule to be

$$v'_{1i} = v_{1i} - V_i^{CM} = \frac{m_2}{m_1 + m_2}(v_{1i} - v_{2i})$$

$$v'_{2i} = v_{2i} - V_i^{CM} = -\frac{m_1}{m_1 + m_2}(v_{1i} - v_{2i}).$$

The velocities v'_{1f} and v'_{2f} after collision, relative to the CM frame, can be determined using the energy and momentum conservation laws:

$$v'_{1f} = -v'_{1i}; \quad v'_{2f} = -v'_{2i}.$$

The other set of solutions, $v'_{1f} = v'_{1i}$ and $v'_{2f} = v'_{2i}$, corresponds to the physically uninteresting case of no collision. Relative to the laboratory frame, the velocities of the two particles after collision are

$$v_1 = \left(\frac{m_1 - m_2}{m_1 + m_2}\right) v_{1i} + \left(\frac{2m_2}{m_1 + m_2}\right) v_{2i} + g \sin \theta t$$

$$v_2 = \left(\frac{2m_1}{m_1 + m_2}\right) v_{1i} - \left(\frac{m_1 - m_2}{m_1 + m_2}\right) v_{2i} + g \sin \theta t.$$

What is the point in introducing the CM frame in this problem? Could we not simply work in the laboratory frame? We could, but that would have required us to calculate the velocities of the bodies just before the collision, which means we would have had to know the instant of collision. The CM frame avoids this difficulty since the velocities relative to this frame are constant except during collision. Note that t in our calculation is the time elapsed from the initial instant $t = 0$ and not from the instant of collision. Of course the instant of collision can be easily obtained by dividing the initial separation by the initial relative velocity.

Are we justified in using energy and momentum conservation in this problem even when the system of colliding bodies is subject to an external force (gravity)? Yes. Since the collision is assumed to take place in an infinitesimal time interval, the external force causes vanishingly small change in the momentum and energy of the system.

P13 *A sealed cylinder full of an incompressible liquid rotates about its axis with a constant angular velocity ω in a gravity-free space. What is the pressure distribution of the liquid in the cylinder?*

What happens to a small colloidal suspension (which is denser than the liquid) in the liquid?

A13 Consider a volume element $dxdydz$ in the liquid observed relative to the rotating rest frame of the cylinder. Since the element is at rest with respect to the frame, it experiences a centrifugal force but no Coriolis force. There must, therefore, be a radial pressure gradient such that the resulting inward force on the element balances the outward centrifugal force.

The centrifugal force is given by $\boldsymbol{F}_{cf} = -m\boldsymbol{\omega} \times (\boldsymbol{\omega} \times \boldsymbol{r})$. With $\boldsymbol{\omega} = \omega\widehat{k}$ and $\boldsymbol{r} = r\widehat{r}$,

$$\boldsymbol{F}_{cf} = \rho\omega^2(x\widehat{i} + y\widehat{j})dxdydz$$

where ρ is the density of the liquid.

The force $\boldsymbol{F}$ on the element due to the pressure gradient is

$$\begin{aligned}\boldsymbol{F} &= [-p(x+dx)dydz + p(x)dydz]\widehat{i} \\ &\quad + [-p(y+dy)dxdz + p(y)dxdz]\widehat{j} \\ &= -\left[\frac{\partial p}{\partial x}\widehat{i} + \frac{\partial p}{\partial y}\widehat{j}\right] dxdydz.\end{aligned}$$

Thus $$\frac{\partial p}{\partial x} = \rho\omega^2 x; \qquad \frac{\partial p}{\partial y} = \rho\omega^2 y; \qquad \frac{\partial p}{\partial z} = 0.$$

Now,

$$\begin{aligned}dp &= \frac{\partial p}{\partial x}dx + \frac{\partial p}{\partial y}dy \\ &= \rho\omega^2(xdx + ydy) = \frac{1}{2}\rho\omega^2 d(r^2)\end{aligned}$$

where $r^2 = x^2 + y^2$. In free space $p(r = 0) = 0$ and ρ is constant for an incompressible liquid. Hence, we obtain

$$p = \frac{1}{2}\rho\omega^2 r^2.$$

Now consider a colloidal suspension in the liquid. Let ρ' be the density of a particle of the suspension and let $dxdydz$ be its volume. The particle experiences a pressure gradient force identical to that experienced by a liquid element of the same size kept in its place; it is proportional to the density ρ of the liquid. The centrifugal force on the particle, however, is proportional to its density ρ'. Hence, if $\rho' > \rho$ the particle experiences a net outward force. On the other hand, if $\rho' < \rho$, the net force is inward.

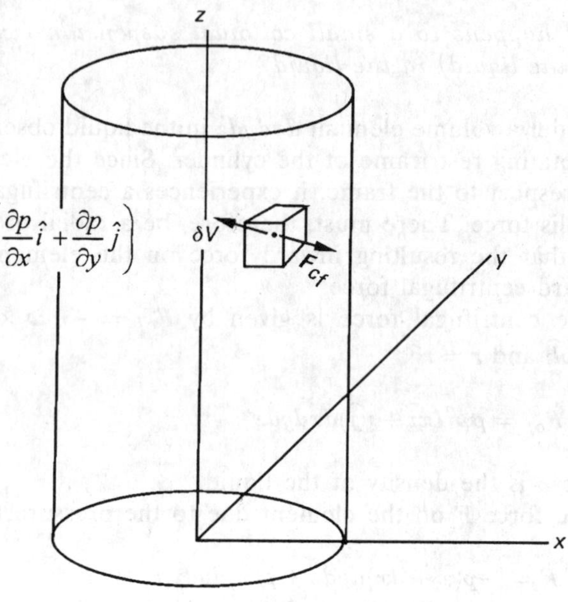

Fig. 9.16

P14 *The shape of a free liquid surface is a plane, neglecting the edge effects. What changes in this shape do you expect if a cylinder containing a liquid is accelerated uniformly in: (a) a vertical direction (b) a horizontal direction? (c) What is the shape if it is rotated about its axis with a uniform angular velocity w?*

A14 Consider a volume element $dxdydz$ in the liquid observed relative to the rest frame of the liquid.

(a) Since the container accelerates in the upward direction, the rest frame of the liquid is a non-inertial frame with upward acceleration, say α_z along the z-axis. The element is in equilibrium in this frame, so that

$$F_x = 0; \quad F_y = 0; \quad F_z - m\alpha_z = 0.$$

Now, the force on the element due to pressure gradient is given by

$$\begin{aligned} F_x &= -\frac{\partial p}{\partial x}dxdydz \\ F_y &= -\frac{\partial p}{\partial y}dxdydz \\ F_z &= -\frac{\partial p}{\partial z}dxdydz. \end{aligned} \tag{A14.1}$$

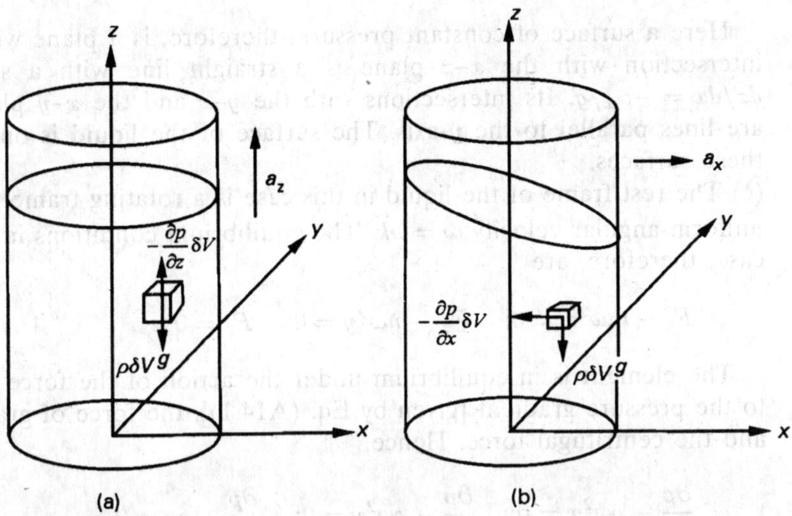

Fig. 9.17

Hence,

$$\frac{\partial p}{\partial x} = 0; \quad \frac{\partial p}{\partial y} = 0; \quad \frac{\partial p}{\partial z} + \rho(\alpha_z + g) = 0,$$

which gives

$$p = p_0 - \rho(\alpha_z + g)(z - z_0).$$

Thus in this case the surfaces of constant p are the surfaces of constant z, i.e. horizontal surfaces. The surface of the liquid is a constant pressure surface (p = atmospheric pressure) and hence, must be horizontal.

(b) In this case, the container accelerates in a horizontal direction. Let the acceleration of the non-inertial rest frame of the liquid be α_x along the x-axis. Equilibrium of the element in this frame implies

$$F_x - m\alpha_x = 0; \quad F_y = 0; \quad F_z = 0.$$

Using Eq. (A14.1), we get

$$\frac{\partial p}{\partial x} + \rho\alpha_x = 0; \qquad \frac{\partial p}{\partial y} = 0; \qquad \frac{\partial p}{\partial z} + \rho g = 0.$$

These equations can be easily solved to obtain

$$p = -\rho(\alpha_x x + gz) + C.$$

Here a surface of constant pressure, therefore, is a plane whose intersection with the x–z plane is a straight line with a slope $dz/dx = -\alpha_x/g$. Its intersections with the y–z and the x–y planes are lines parallel to the y-axis. The surface of the liquid is one of these surfaces.

(c) The rest frame of the liquid in this case is a rotating frame with uniform angular velocity $\boldsymbol{\omega} = \omega\hat{k}$. The equilibrium conditions in this case, therefore, are

$$F_x - m\omega^2 x = 0; \quad F_y - m\omega^2 y = 0; \quad F_z = 0.$$

The element is in equilibrium under the action of the force due to the pressure gradient [given by Eq. (A14.1)], the force of gravity and the centrifugal force. Hence,

$$\frac{\partial p}{\partial x} - \rho\omega^2 x = 0; \quad \frac{\partial p}{\partial y} - \rho\omega^2 y = 0; \quad \frac{\partial p}{\partial z} + \rho g = 0.$$

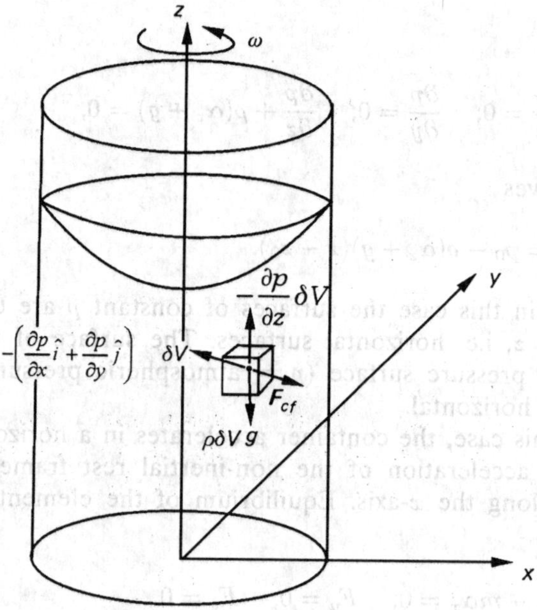

Fig. 9.18

The solution of this equation is

$$p = \rho\left(\frac{1}{2}\omega^2 r^2 - gz\right) + p_0$$

where p_0 is the pressure at $r = 0$, $z = 0$. The constant pressure surface with p = atmospheric pressure, i.e. the surface of the liquid, is thus a paraboloid of revolution. This is something you know from Newton's pail experiment.

P15 *A stone of mass m is tied to a string and is rotated in a vertical circle. Calculate the tension in the string if it is just able to move the stone along the circle.*

A15 Let us take the plane of motion of the stone to be the x–y plane in the laboratory frame, with the y-axis along the vertical and the z-axis along the direction of ω (see Fig. 9.19). The centripetal force for the circular motion is provided by the tension in the string and the component of weight along the string:

$$mR\omega^2 = T + mg\sin\theta \tag{A15.1}$$

where R is the length of the string. The angular acceleration is determined by the torque, $\tau = -mgR\cos\theta$ of the weight of the stone about the centre. The moment of inertia of the stone about the centre is mR^2. Therefore

$$mR^2\frac{d\omega}{dt} = -mg \times R\cos\theta,$$

i.e.
$$\frac{d\omega}{dt} + \frac{g}{R}\cos\theta = 0. \tag{A15.2}$$

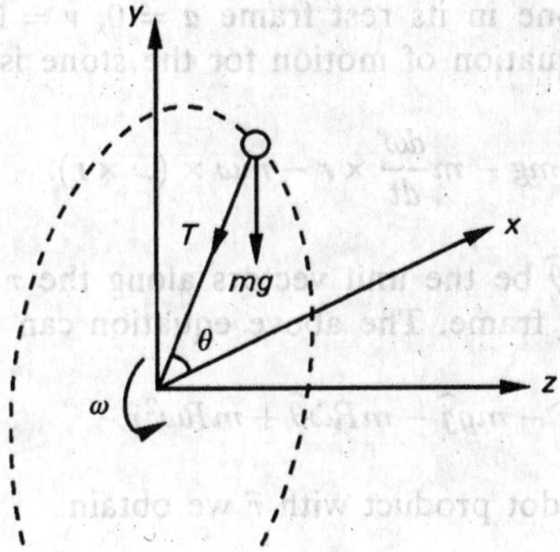

Fig. 9.19

Note that the tension in the string is not constant since in Eq. (A15.1) both ω and θ vary. It is minimum at the highest point. If the stone is just able to move in circle, this minimum value is zero, i.e. at $\theta = \pi/2$, $T = 0$, and hence $\omega = \sqrt{(g/R)}$. Multiplying both the sides of Eq. (A15.2) by ωdt we can rewrite this equation as

$$d\left(\frac{1}{2}\omega^2 + \frac{g}{R}\sin\theta\right) = 0.$$

Hence,

$$\frac{1}{2}\omega^2 + \frac{g}{R}\sin\theta = \textit{constant}.$$

Since $\omega = \sqrt{(g/R)}$ at $\theta = \pi/2$, the constant above is $3g/2R$. We thus obtain

$$\omega = \sqrt{\frac{2g}{R}\left(\frac{3}{2} - \sin\theta\right)}$$

$$T = 3mg(1 - \sin\theta).$$

It is instructive to look at the problem from the rotating rest frame of the particle. The equation of motion in a rotating frame is

$$m\boldsymbol{a} = \boldsymbol{F} - m\frac{d\omega}{dt} \times \boldsymbol{r} - 2m\boldsymbol{\omega} \times \boldsymbol{v} - m\omega \times (\omega \times \boldsymbol{r}).$$

For the stone in its rest frame $\boldsymbol{a} = 0$, $\boldsymbol{v} = 0$ and $\boldsymbol{F} = \boldsymbol{T} + m\boldsymbol{g}$. Hence, the equation of motion for the stone is

$$0 = \boldsymbol{T} + m\boldsymbol{g} - m\frac{d\omega}{dt} \times \boldsymbol{r} - m\omega \times (\omega \times \boldsymbol{r}).$$

Let $\widehat{r}$ and $\widehat{\theta}$ be the unit vectors along the r and θ directions in the laboratory frame. The above equation can thus be written as

$$0 = -T\widehat{r} - mg\widehat{j} - mR\dot{\omega}\widehat{\theta} + mR\omega^2\widehat{r}.$$

Taking the dot product with $\widehat{r}$ we obtain

$$T = -mg\sin\theta + mR\omega^2$$

which is identical to Eq. (A15.1). And the dot product with $\widehat{\theta}$ gives

$$\dot{\omega} = -\frac{g\cos\theta}{R},$$

which is identical to Eq. (A15.2).

This problem involves a rotating frame with a non-uniform angular velocity. Most theoretical treatments of non-inertial frames consider uniformly rotating frames and therefore include only the centrifugal and Coriolis forces. The existence of the term $-m d\boldsymbol{\omega}/dt \times \boldsymbol{r}$ in Newton's II law for a rotating frame, therefore, goes unnoticed. As we have seen, this term plays a crucial role in the above problem.

Rotational Dynamics

P16 *ABC is a triangular lamina of uniform mass density, its total mass being M. Show that the moment of inertia of the lamina about any axis in its plane and passing through one of its vertices is $(M/6)(\alpha^2 + \beta^2 + \alpha\beta)$, where α and β are the perpendicular distances of the other two vertices from the axis.*

A16 The moment of inertia about an axis, $I = \int dmr^2$, where r is the distance of the elemental mass dm from the axis and the integration is to be performed over the entire body. This is possible when the body has a regular shape and the axis happens to be conveniently oriented. In general, however, the integration involved is hard if not impossible. In this problem we illustrate a simple fact about the moment of inertia which is often useful: the moment of inertia of a body about an axis is equal to the sum of the moments of inertia of its individual parts about the same axis.

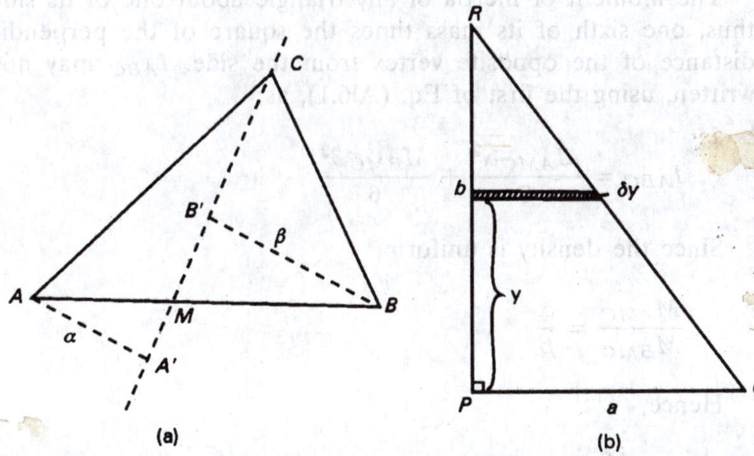

Fig. 9.20

For the triangular lamina of Fig. 9.20(a) and a given axis, we have

$$I_{ABC} = I_{AMC} + I_{BMC};$$

$$I_{AMC} = I_{AA'C} - I_{AA'M} \qquad \text{(A16.1)}$$

$$I_{BMC} = I_{BB'C} + I_{BB'M}.$$

The problem, therefore, reduces to finding the I of a right-angled triangular lamina about one of its sides. This can be done explicitly. For the right-angled triangle PQR in Fig. 9.20(b),

$$I = \int dmy^2 = \rho \int dxdyy^2.$$

From the figure it is seen that, for a given y, x goes from 0 to $a(b-y)/b$ and the range of y is from 0 to b. Thus

$$I = \rho \int_0^b dyy^2 \int_0^{\frac{a}{b}(b-y)} dx = \rho \frac{ab^3}{12}$$
$$= \frac{Mb^2}{6}.$$

Using this result in the last two of Eq. (A16.1), we get

$$I_{AMC} = \frac{M_{AMC}\alpha^2}{6}; \quad I_{BMC} = \frac{M_{BMC}\beta^2}{6}.$$

The moment of inertia of any triangle about one of its sides is, thus, one sixth of its mass times the square of the perpendicular distance of the opposite vertex from the side. I_{ABC} may now be written, using the first of Eq. (Al6.1), as

$$I_{ABC} = \frac{M_{AMC}\alpha^2}{6} + \frac{M_{BMC}\beta^2}{6}.$$

Since the density is uniform,

$$\frac{M_{AMC}}{M_{BMC}} = \frac{\alpha}{\beta}.$$

Hence,

$$I = \frac{M}{6}(\alpha^2 + \beta^2 + \alpha\beta).$$

P17 *A car of mass M is moving with a speed v_0 along a straight road. It is brought to rest over a distance L by applying the brakes. Assume that the wheels stop rotating completely as soon as the brakes are applied and the car slides to rest. Determine the coefficient of kinetic friction μ_k between the tyres and the road and the reaction forces by the ground at each of the tyres. As shown in Fig. 9.21, the distance between the front and the rear wheels is 2ℓ and the CM of the car is midway between the wheels at a height h above the ground.*

A17 The free-body diagram of the car is shown in Fig. 9.21. N_1, f_1 are the normal force and the frictional force at the front wheels and N_2, f_2 are those at the rear wheels. Mg, the weight of the car, acts at its CM.

Since the initial velocity of the car is v_0 and the distance travelled before coming to rest is L, the horizontal retardation of the car has magnitude $a = v_0^2/2L$. Hence,

$$\begin{aligned} Ma &= f_1 + f_2 \\ &= \mu_k(N_1 + N_2). \end{aligned}$$

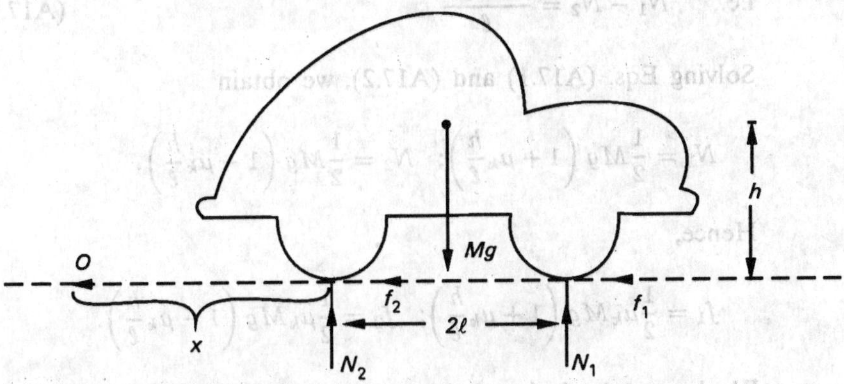

Fig. 9.21

The direction of the force of kinetic friction is opposite to the motion of the car. There is no motion in the vertical direction; therefore

$$N_1 + N_2 = Mg. \tag{A17.1}$$

Clearly,

$$\mu_k = \frac{a}{g} = \frac{v_0^2}{2Lg}.$$

These equations do not determine N_1 and N_2 separately. We have so far considered only the translational motion which determines the sum $N_1 + N_2$. To determine these forces separately we need one more equation and that is provided by the angular momentum consideration. Naively one might expect N_1 to be equal to N_2, since there is an apparent symmetry between the front and the rear wheels. As we shall see, this expectation is incorrect because the car is accelerated. Let us employ the rotational law of motion in the CM frame:

$$\frac{d\mathbf{L}'_{CM}}{dt} = \mathbf{N}'_{CM}.$$

Recall that this law has the same form as in the laboratory frame, despite the fact that the CM frame is accelerated. Obviously, $\mathbf{L}'_{CM}$ and $d\mathbf{L}'_{CM}/dt$ are zero as there is no motion relative to CM since the wheels have stopped rotating. So, the net torque about the CM must be zero:

$$(N_1 - N_2) \times \ell - \mu_k(N_1 + N_2) \times h = 0,$$

$$\text{i.e.,} \quad N_1 - N_2 = \frac{\mu_k Mgh}{\ell}. \tag{A17.2}$$

Solving Eqs. (A17.1) and (A17.2), we obtain

$$N_1 = \frac{1}{2}Mg\left(1 + \mu_k\frac{h}{\ell}\right); \quad N_2 = \frac{1}{2}Mg\left(1 - \mu_k\frac{h}{\ell}\right).$$

Hence,

$$f_1 = \frac{1}{2}\mu_k Mg\left(1 + \mu_k\frac{h}{\ell}\right); \quad f_2 = \frac{1}{2}\mu_k Mg\left(1 - \mu_k\frac{h}{\ell}\right).$$

We have solved the problem using the rotational equilibrium condition relative to the CM frame. But it is not necessary to go to the CM frame. Let us employ the angular momentum equation relative to the ground frame:

$$\frac{d\mathbf{L}_O}{dt} = \mathbf{N}_O.$$

O is chosen to be the initial position of the rear wheel. The net torque about O is

$$\begin{aligned} N &= N_2 \times x + N_1 \times (x + 2\ell) - Mg \times (x + \ell) \\ &= (N_1 - N_2)\ell. \end{aligned}$$

The angular momentum of the car about O is $L_O = -Mvh.$ Hence,

$$\frac{dL_O}{dt} = Mah.$$

Equating the two, we get Eq. (A17.2).

Although the acceleration of the car spoils the symmetry between N_1 and N_2, there is an obvious symmetry between the two wheels on the rear side as also between the wheels on the front side. Therefore there is a reaction force $N_1/2$ at each of the front wheels and $N_2/2$ at each of the rear wheels.

P18 *A block of mass m_2 is pulled along the top of a table with a rope passing over a pulley as shown in Fig. 9.22(a). Another mass m_1 is attached to the other end of the rope. The mass of the pulley is M and its radius R. Assume that there is no slippage between the rope (of negligible mass) and the pulley. Calculate the acceleration of the system.*

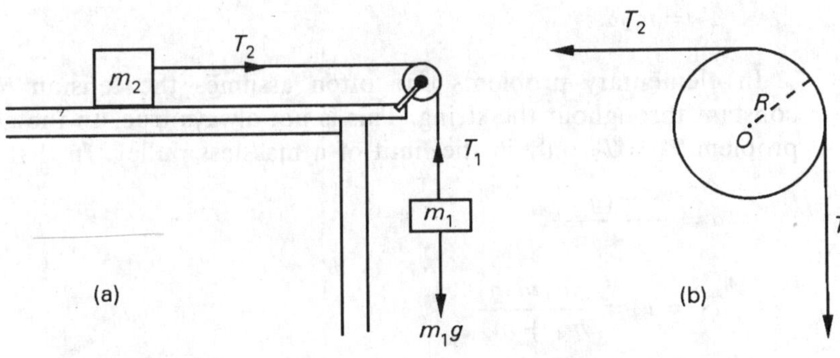

Fig. 9.22

A18 Since there is no slippage between the rope and the pulley, the angular acceleration α of the pulley and the linear acceleration a of the two masses are related,

$$a = R\alpha.$$

As the masses accelerate, the angular velocity of the pulley increases. Clearly, there must be a net torque about the axis of the pulley. The reaction of the table acting at O has zero torque about O. The torque due to tension in the string is $R \times (T_1 - T_2)$ [see Fig. 9.22(b)]. † T_1 and T_2 cannot, therefore, be equal. Since

† This needs an explanation, since T_1. and T_2 are forces at the two ends of the

the moment of inertia of the pulley about its axis is $1/2MR^2$, the equation of motion for the pulley is

$$(T_1 - T_2) \times R = \frac{1}{2}MR^2\alpha.$$

The equations of motion for m_1 and m_2 are

$$m_1 a = m_1 g - T_1$$

$$m_2 a = T_2.$$

The four equations can be solved to obtain

$$\alpha = \frac{m_1 g}{(m_1 + m_2 + \frac{M}{2})R}$$

$$a = R\alpha$$

$$T_1 = m_1(g - a)$$

$$T_2 = m_2 a.$$

In elementary problems one often assumes the tension to be constant throughout the string. This is not always true. In the above problem $T_1 = T_2$ only in the limit of a massless pulley. In that case

$$a = \frac{m_1 g}{m_1 + m_2}$$

$$T_1 = T_2 = \frac{m_1 m_2 g}{m_1 + m_2}.$$

These familiar results are also obtained if we consider a massive but clamped pulley, over which the string slips without friction. In this case the pulley has no role to play in the motion of the system except in changing the direction of tension. Note also that, whatever the situation, tension in a massive accelerated string is not uniform (see T38, Ch.4).

---

segment of the string in contact with the pulley. How can we equate them to the forces on the pulley? Consider an infinitesimal element of this segment as the free-body. There must be a force of static friction δf on this element which prevents slippage. The net force on the element is zero since the element, though accelerated, is mass less. Hence, δ_f must be balanced by the difference in tension δT at the two ends of the element. By the III law the force on the pulley by the element is $-\delta f = \delta T$. Each element produces a torque $R \times \delta T$ on the pulley about its axis. Hence, the total torque on the pulley is $R \times (T_1 - T_2)$.

P19 *A disc of mass M and radius R in a vertical plane rotates about its axis with a constant angular velocity ω_1. The axle is mounted on a vertical frame, as shown in Fig. 9.23, which rotates about the vertical with a constant angular velocity ω_2. What is the torque on the disc?*

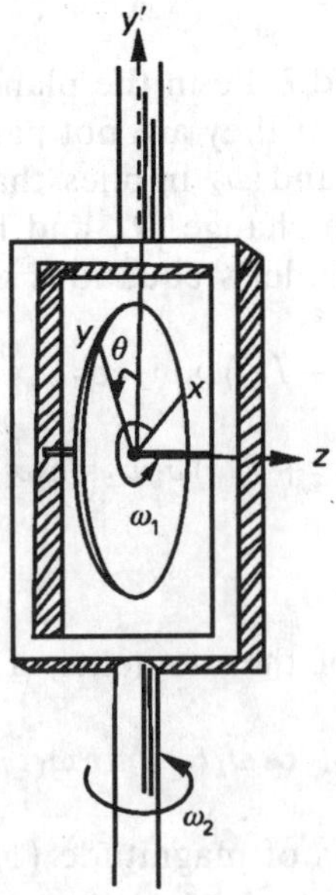

Fig. 9.23

A19 The principal axes of the disc are easy to locate. Any two mutually perpendicular diameters of the disc and the axis of the disc form a set of principal axes. The angular velocity ω of the disc is in general not parallel to any of the principal axes; it has non-zero components along the axis of the disc and the instantaneous vertical diameter. We must, therefore, use Euler's equations of rigid body motion.

Let us choose a set of body-fixed axes along principal axes as follows: y-axis along the initially vertical diameter, z-axis along the axis of the disc and x-axis such that the three form a right-handed triad. For the laboratory frame we take the y'-axis along the vertical, the z'-axis along the initial position of the z-axis (i.e. the axis of the disc) and the x'-axis to form a right-handed triad.

Relative to the body-fixed frame, the moment of inertia tensor is diagonal with $I_{xx} = I_{yy} = (1/4)MR^2$ and $I_{zz} = (1/2)MR^2$. The angular velocity of the disc is

$$\omega = \omega_2 \hat{j}' + \omega_1 \hat{k}$$
$$= \omega_2(\sin\omega_1 t\hat{i} + \cos\omega_1 t\hat{j}) + \omega_1 \hat{k}$$

where we have used the fact that $\hat{j}' = \sin\omega_1 t\hat{i} + \cos\omega_1 t\hat{j}$ as can be seen from the figure. The angular momentum of the disc is

$$\begin{aligned} \boldsymbol{L} &= I_{xx}\omega_2 \sin\omega_1 t\hat{i} + I_{yy}\omega_2\cos\omega_1 t\hat{j} + I_{zz}\omega_1\hat{k} \\ &= \frac{1}{4}MR^2\omega_2\hat{j}' + \frac{1}{2}MR^2\omega_1\hat{k}. \end{aligned}$$

Clearly, both $\boldsymbol{\omega}$ and $\boldsymbol{L}$ lie in the plane containing the vertical and the axis of the disc, but they are not parallel to each other. Further, the constancy of ω_1 and ω_2 implies that they are both of constant magnitudes. Thus the change $\delta\boldsymbol{L}$, and hence the torque $\boldsymbol{\tau}$, must be perpendicular to $\boldsymbol{L}$. Euler's equations give

$$\tau_x = (I_{xx} - I_{yy} + I_{zz})\omega_1\omega_2\cos\omega_1 t$$

$$\tau_y = (-I_{yy} - I_{zz} + I_{xx})\omega_1\omega_2\sin\omega_1 t$$

$$\tau_z = 0.$$

Using the values of the moments of inertia we obtain

$$\boldsymbol{\tau} = \tfrac{1}{2}MR^2\omega_1\omega_2(\cos\omega_1 t\hat{i} - \sin\omega_1 t\hat{j})$$

The torque, thus, is of magnitude $(1/2)MR^2\omega_1\omega_2$ and acts along the horizontal direction $\hat{j}'\times\hat{k}$ in the plane of the disc; it is perpendicular to $\boldsymbol{L}$ as expected.

This problem illustrates the points made in T28, Chapter 8. It is an example of rotation about a fixed point but not a fixed axis since the direction of $\boldsymbol{\omega}$ changes with time. Further the angular momentum and angular velocity point in different directions so that the relation $L = I\omega$ that one uses in elementary mechanics is not valid in this case. By the same token the relation $\boldsymbol{\tau} = I\boldsymbol{\alpha}$ is also not valid and one needs to invoke the general Euler's equations of motion.

P20 *A constant torque $\boldsymbol{\tau}$ is applied to the disc of the previous problem, parallel to the axle (e.g. by applying a constant tension in a string wound on the rim of the disc). If the initial angular velocity of the disc is ω_0 about the vertical axis, calculate the angular velocity as a function of time.*

A20 Relative to the body-fixed frame, the torque $\boldsymbol{\tau} = \tau\hat{k}$. The Euler's equations, therefore, reduce to

$$\tau = I_{zz}\dot{\omega}_z = \tfrac{1}{2}MR^2\dot{\omega}_z$$

$$0 = I_{xx}\dot{\omega}_x - (I_{yy} - I_{zz})\omega_y\omega_z = \dot{\omega}_x + \omega_y\omega_z \tag{A20.1}$$

$$0 = I_{yy}\dot{\omega}_y - (I_{zz} - I_{xx})\omega_z\omega_x = \dot{\omega}_y - \omega_z\omega_x.$$

The first of these equations gives

$$\omega_z = \frac{\tau}{\frac{1}{2}MR^2}t$$

$$\theta = \frac{\tau}{\frac{1}{2}MR^2}\frac{t^2}{2}$$

where we have assumed the initial conditions: $\theta = 0$ and $\omega_z = 0$ at $t = 0$.

In order to solve the remaining two parts of Eq. (A20.1), we take their ratio and rewrite the resulting equation as

$$\frac{1}{2}\frac{d}{dt}(\omega_x^2 + \omega_y^2) = 0.$$

Hence,

$$\omega_x^2 + \omega_y^2 = \text{constant}.$$

Now, the initial velocity of the disc is $\omega_0\hat{j}'$. But at $t = 0$ the y- and y'-axes coincide. Hence, $\omega = \omega_0\hat{j}$ at $t = 0$ and the constant in the above equation is ω_0^2, so

$$\omega_x^2 + \omega_y^2 = \omega_0^2.$$

We may now write the second part of Eq. (A20.1) as

$$\frac{d\omega_x}{dt} = \sqrt{\omega_0^2 - \omega_x^2}\,\frac{\tau}{I_{zz}}t.$$

Integrating this equation with the initial condition $\omega_x = 0$ at $t = 0$, we obtain

$$\omega_x = \omega_0 \sin\theta$$

and hence,

$$\omega_y = \omega_0 \cos\theta.$$

Thus the angular velocity is

$$\omega = \omega_0 \sin\theta\hat{i} + \omega_0 \cos\theta\hat{j} + \left(\frac{\tau}{I_{zz}}\right) t\hat{k}$$
$$= \omega_0\hat{j}' + \left(\frac{\tau}{I_{zz}}\right) t\hat{k}.$$

The vertical component of the angular velocity remains unaffected by the torque, which merely increases the angular velocity of the disc about its axis. One could have expected this intuitively. However, this decoupling of the components of ω is not as obvious as one might think. It is possible only because of the symmetry $(I_{xx} = I_{yy})$ in the problem. An elliptical disc in the same situation will not show this decoupling.

Work–Energy Theorem

P21 *A disc of radius R and mass M, rotating with an angular velocity ω_0, is placed lightly (i.e. without imparting any translational velocity to its CM) on a horizontal table. The coefficient of kinetic friction between the disc and the table is μ_k. After how long does the disc start rolling? Describe the energetics of the motion of the disc.*

A21 Initially, the velocity v_{CM} of the CM of the disc is zero and its initial angular velocity is ω_0. Since the contact between the disc and the table is rough, the frictional force opposes relative motion between the two surfaces at the point of contact. The ensuing motion of the disc is rolling with slippage. The sense of rotation of the disc and the direction of the frictional force f on the disc are shown in Fig. 9.24. Clearly, f accelerates the CM of the disc while the torque due to f retards its angular velocity. When v_{CM} and ω reach such values that $v_{CM} = R\omega$ is satisfied, the disc starts rolling without slipping.

The force of friction $f = \mu_k$Mg. Hence, the translational acceleration $a = \mu_k g$ and v_{CM} is given by

$$v_{CM} = \mu_k g t.$$

The torque due to friction is $\tau = \mu_k MgR$. Since $I = (1/2)MR^2$. the angular acceleration is $\alpha = -2\mu_k g/R$. Therefore

$$\omega = \omega_0 - \frac{2\mu_k g}{R} t.$$

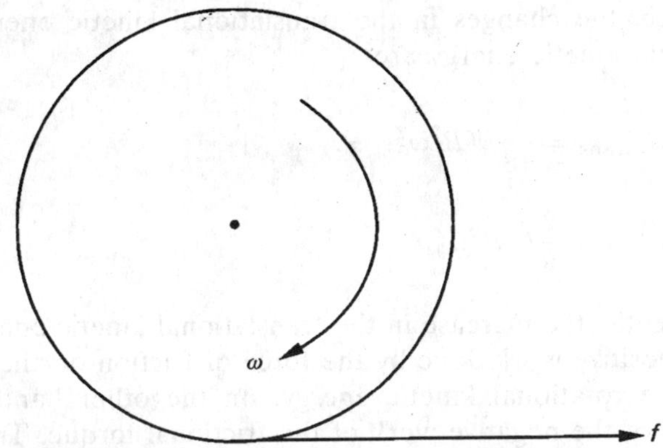

Fig. 9.24

The rolling condition $v_{CM} = R\omega$ will be satisfied at

$$T = \frac{\omega_0 R}{3\mu_k g}.$$

Let us now calculate the work done during this time by the force of friction in displacing the CM of the disc:

$$W_f = \int_0^T f v_{\text{CM}} dt = (\mu_k g)^2 M \int_0^T t dt$$
$$= \frac{1}{18} M R^2 \omega_0^2.$$

The work done by the frictional torque is

$$W_\tau = -\int_0^T \tau \omega dt = -\mu_k g M R \int_0^T \left(\omega_0 - \frac{2\mu_k g t}{R}\right) dt$$
$$= -\frac{2}{9} M R^2 \omega_o^2.$$

The total work done by friction, thus, is

$$W = -\frac{3}{18} M R^2 \omega_0^2.$$

Now, the velocity of the CM $v_{CM,f}$ and the angular velocity ω_f at $t = T$ are

$$v_{CM,f} = \frac{\omega_0 R}{3};\ \omega_f = \frac{\omega_0}{3}.$$

Hence, the changes in the translational kinetic energy and the rotational kinetic energy are

$$\Delta K_{\text{trans}} = \frac{1}{18} M R^2 \omega_0^2,$$

$$\Delta K_{\text{rot}} = \frac{2}{9} M R^2 \omega_0^2.$$

Evidently, the increase in the translational kinetic energy is equal to the positive work done by the force of friction on the CM of the disc. The rotational kinetic energy, on the other hand, decreases because of the negative work of the frictional torque. The net work done by friction is negative and, accordingly, there is a net loss in the kinetic energy of the disc.

Note the interesting lesson we learn from this problem. Ordinarily, we think of friction as something that retards motion. Here, instead, friction is responsible for moving the disc on the plane. (Without friction the disc would simply keep rotating at the same point.) Overall, however, kinetic friction is necessarily dissipative which is why there is a loss of kinetic energy as we saw above. We have earlier encountered situations where friction causes motion. For example, a box at rest on an accelerated train (see T46 Ch.5) moves because of friction. Other examples are: walking, vehicle negotiating a bend, etc. In these situations, however, it is the static friction that comes into play and there is no dissipation of energy.

Once the disc starts rolling the point of contact is stationary, i.e. the backward velocity ωR is equal and opposite to the forward translational velocity v_{CM}. There is no role for friction anymore and it becomes zero. What if now a torque is applied to increase the angular velocity of the disc? (This is what happens when the accelerator of a car is pressed.) The backward velocity will tend to increase and there will be an impending slippage motion. To prevent this, the force of static friction will come into play. This force causes the acceleration of the CM while the torque increases the angular velocity. Thus the rolling condition will continue to be satisfied. Of course all this presumes that there is sufficient friction, in the absence of which the disc will skid.

P22 *Two particles of equal masses are rotating about their CM at a constant angular velocity ω_0. An additional mutual radial force starts drawing them towards each other. Verify the work–energy theorem.*

A22 Let the initial separation between the two masses be $2R$. By symmetry their radial distances from the CM must be equal at all times.

Let us take the origin at the CM and use plane polar co-ordinates. The instantaneous positions of the two masses can be written as

$$\boldsymbol{r}_1 = r\widehat{r}(\theta_1); \quad \boldsymbol{r}_2 = r\widehat{r}(\theta_2)$$

where $\theta_2 = \theta_1 + \pi$. This is always so because the forces are internal so that the CM is stationary. The velocities of the two masses are given by

$$\boldsymbol{v}_1 = \dot{r}\widehat{r}(\theta_1) + r\omega\widehat{\theta}(\theta_1) \tag{A22.1}$$

$$\boldsymbol{v}_2 = \dot{r}\widehat{r}(\theta_2) + r\omega\widehat{\theta}(\theta_2)$$

where we have used the fact that

$$\dot{\theta}_1 = \dot{\theta}_2 = \omega.$$

The angular momentum of the system is

$$\boldsymbol{L} = 2mr^2\omega\widehat{k} \tag{A22.2}$$

where $\widehat{k} = \widehat{r} \times \widehat{\theta}$. Since the net torque on the system is zero, $\boldsymbol{L}$ is conserved. Clearly, as r becomes smaller, the angular velocity increases. The kinetic energy of the system is

$$K = 2 \times \frac{1}{2}m(\dot{r}^2 + r^2\omega^2) = m\dot{r}^2 + \frac{L^2}{4mr^2}.$$

Hence, the rate at which the kinetic energy increases is given by

$$\frac{dK}{dt} = 2m\dot{r}\ddot{r} - \frac{L^2\dot{r}}{2mr^3}. \tag{A22.3}$$

By the work–energy theorem this change must be accounted for by the work of the radial force. The expression for force is

$$\boldsymbol{F} = m[(\ddot{r} - r\omega^2)\widehat{r} + (r\dot{\omega} + 2\dot{r}\omega)\widehat{\theta}].$$

In the present problem, $\boldsymbol{F}$ is given to be radial; thus the θ component of $\boldsymbol{F}$ is zero. The forces on the two particles, using Eq. (A22.2), are

$$\boldsymbol{F}_1 = m\left(\ddot{r} - \frac{L^2}{4m^2r^3}\right)\widehat{r}(\theta_1); \; \boldsymbol{F}_2 = m\left(\ddot{r} - \frac{L^2}{4m^2r^3}\right)\widehat{r}(\theta_2) \tag{A22.4}$$

The total power delivered by the forces, $\boldsymbol{F}_1 \cdot \boldsymbol{v}_1 + \boldsymbol{F}_2 \cdot v_2$, can be easily obtained using Eqs. (A22.1) and (A22.4). It is equal to the rate of increase of kinetic energy given by Eq. (A22.3).

Note a possible pitfall. Initially the particles are subject to a radial force but no radial motion, i.e. the force does no work and the kinetic energy is constant. When the extra radial force appears, radial motion ensues. It will be wrong to think that the change in energy is brought about by the work due to the extra force only. As we have seen above, the work done by the total radial force equals the change in kinetic energy.

This problem in a way models a familiar situation that is often used to illustrate angular momentum conservation: an acrobat standing on a rotating platform pulling her outstretched hands inward and thereby increasing her angular velocity. The rotational kinetic energy increases due to the positive work done by the internal forces exerted by the acrobat in pulling the arms inward. What if she stretches the arms outward again? The angular velocity will decrease by angular momentum conservation. The kinetic energy will decrease since work done by the internal forces is now negative. This is because the centripetal force necessary to keep her rotating is inward while the displacement is outward. (The arms are stretched out by slightly reducing the centripetal force on the arms provided by the rest of the body.)

Collision with an extended body

P23 *B is an assembly of a light rod of length 2ℓ with two equal masses at its ends. A point object A of mass m_A moving normal to the rod gets scattered off one of the masses at an angle θ with its original direction of motion. For the case when $m_A = m_B$ and $\theta = \pi/6$, calculate what fraction of the initial kinetic energy is retained by A assuming the collision to be elastic.*

A23 This is a straightforward problem of elastic collision. There is, however, one important thing to note. It is a collision involving an extended body. Therefore, besides the conservation of linear momentum and kinetic energy (which we invoke for collision between point masses), we will need to invoke the conservation of angular momentum also. The latter will give us an independent equation.

Using conservation of the components of the linear momentum along and transverse to the direction of motion, we get

$$m_A v_{Ai} = m_A v_{Af} \cos\theta + m_B v_{Bf} \cos\phi \qquad \text{(A23.1)}$$

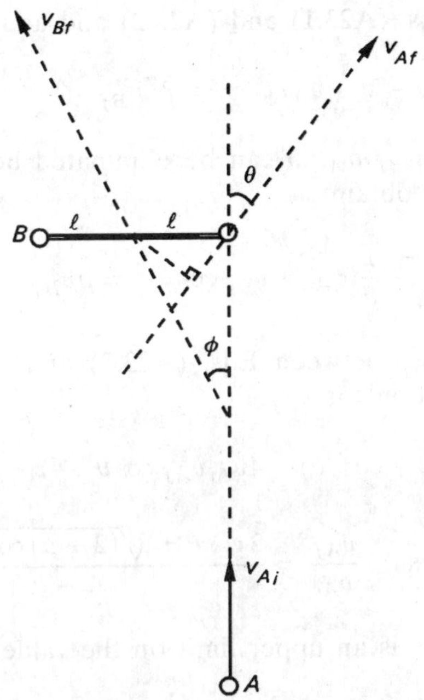

Fig. 9.25

and $\quad m_A v_{Af} \sin\theta = m_B v_{Bf} \sin\phi \qquad$ (A23.2)

where v_{Af} is the velocity of A and v_{Bf} is the velocity of the CM of B after collision.

Let us now employ conservation of angular momentum in the laboratory frame. Choose the initial position O of the centre of mass of B to be the point about which to take moments. The angular momentum of B about O is equal to its angular momentum about the CM in the CM frame plus the angular momentum of B about O assuming all its mass is concentrated at the CM. This latter quantity is zero since the change in velocity of the CM is impulsive, i.e. v_{Bf} passes through O. We can, therefore, write the equation for conservation of angular momentum as

$$m_A v_{Ai} \times \ell = m_A v_{Af} \times \ell \cos\theta + I\omega \qquad \text{(A23.3)}$$

where $I = m_B \ell^2$ is the moment of inertia of B around its CM.

Finally. conservation of energy gives

$$\frac{1}{2} m_A v_{Ai}^2 = \frac{1}{2} m_A v_{Af}^2 + \frac{1}{2} m_B v_{Bf}^2 + \frac{1}{2} I\omega^2. \qquad \text{(A23.4)}$$

Squaring Eqs. (A23.1) and (A23.2) and adding we get

$$v_{Ai}^2 + v_{Af}^2 - 2v_{Ai}v_{Af}\cos\theta = \mu^2 v_{Bf}^2 \tag{A23.5}$$

where $\mu = m_B/m_A$. ω can be eliminated between Eqs. (A23.3) and (A23.4) to obtain

$$v_{Ai}^2 - v_{Af}^2 - \frac{1}{\mu}(v_{Ai} - v_{Af}\cos\theta)^2 = \mu v_{Bf}^2. \tag{A23.6}$$

Elimating v_{Bf} between Eqs. (A23.5) and (A23.6) we obtain a quadratic equation for v_{Af},

$$v_{Af}^2(1 + \mu + \cos^2\theta) - 4v_{Ai}v_{Af}\cos\theta + (2 - \mu)v_{Ai}^2 = 0. \tag{A23.7}$$

Its solution is $\dfrac{v_{Af}}{v_{Ai}} = \dfrac{2\cos\theta \pm \sqrt{(2+\mu)\cos^2\theta - (2 + \mu - \mu^2)}}{1 + \mu + \cos^2\theta}$.

Clearly, there is an upper limit on the value of θ determined by

$$\cos^2\theta \geq \frac{2 + \mu - \mu^2}{2 + \mu}.$$

If $m_A = m_B$, $\mu = 1$ and the solution simplifies to

$$\frac{v_{Af}}{v_{Ai}} = \frac{2\cos\theta \pm \sqrt{3\cos^2\theta - 2}}{(2 + \cos^2\theta)}$$

and $\cos\theta \geq \sqrt{(2/3)}$. For $\theta = \pi/6$, which satisfies the inequality,

$$\frac{v_{Af}}{v_{Ai}} = \frac{4\sqrt{3} \pm 2}{11}.$$

The fraction T_A of the initial total energy appearing as the translational kinetic energy of A after collision is

$$T_A = \frac{\frac{1}{2}m_A v_{Af}^2}{\frac{1}{2}m_A v_{Ai}^2} = \frac{52 \pm 16\sqrt{3}}{121}.$$

The fractional energy T_B appearing as the kinetic energy of B is

$$T_B = \frac{\frac{1}{2}m_B v_{Bf}^2}{\frac{1}{2}m_A v_{Ai}^2} = \frac{41 \mp 6\sqrt{3}}{121}.$$

The angular velocity of B after collision is given by Eq. (A21.3),

$$\frac{\ell\omega}{v_{Ai}} = \frac{5 \mp \sqrt{3}}{11}$$

Hence, the fractional energy R_B appearing as the rotational kinetic energy of B is

$$R_B = \frac{\frac{1}{2}I\omega^2}{\frac{1}{2}m_A v_{Ai}^2} = \frac{28 \mp 10\sqrt{3}}{121}$$

As expected, $T_A + T_B + R_B = 1$. Thus in this collision involving an extended body, the final energy of the system is not only in the translational mode but also in the rotational mode. The distribution of energy in the two modes is governed by both linear and angular momentum conservation.

Index

acceleration, 92, 214
additive quantum number, 176
adiabatic, 205
air resistance, 90
algebra of scalars, 24
algebra of vectors, 24
Ampere's law, 7
angular momentum vector, 49
angular speed, 132
angular velocity, 125, 196
anisotropic crystals, 33
antineutrino, 178
approximations in physics, 10
Archimedes principle, 108
Aristotle, 90
atmospheric pressure, 268
average acceleration, 54
average of a quantity, 70
average velocity, 54, 77
Avogadro's number, 3, 215
axial vector, 49
axis of symmetry, 225
baryon number conservation, 176
base dimensions, 1
basis vectors, 127
binding energy, 182, 183
binomial expansion, 13
body-fixed frame, 218
bound vector, 27
causality, 101
centre of mass, 95, 189
centrifugal force, 130
centripetal acceleration, 139
charge conservation, 176
charge density, 38
circular motion, 141, 218, 224
clamped pulley, 277
concurrent forces, 187
conservation of energy, 286
conservation of linear momentum, 172
conservative force, 153, 158
conserved vector quantity, 42, 43
contact force, 101, 172
Coriolis force, 130, 239, 141
Coulomb force, 101
covariant equations, 44
cross-product, 30, 186
degrees of freedom, 215
dimensional analysis, 5, 16
dimensionless quantity, 3, 5, 11
dipole moment, 46
Dirac, P A M, 20
dot product, 239
dyadic, 31
dyne, 2
elastic and inelastic collision, 173, 285
elastic spring force, 159
electric dipole moment, 231
electric field, 27, 36
electric susceptibility, 33
electrostatic force, 21, 94, 163
electrostatic potential energy, 164
element of area, 34
equation of motion, 94, 99
equilibrium thermodynamics, 155
equilibrium, 104
equivalence of frames, 47
ether, 144
Euler's theorem, 215
exothermic reaction, 183
external force problems, 170
form invariant, 44
four-component object, 52
four vector, 50, 52
frame of absolute rest, 144
frame of reference, 114
frame-dependent force, 130
free fall, 134

free vectors, 27
freebody diagram, 106
friction, 90, 138
frictionless loop-the-loop apparatus, 168
frictionless pulley, 105
fundamental constants, 20
Galilean transformations, 121
Galileo, 90
gamma ray, 181
gas constant, 17
gas law, 3
Gaussian CGS system, 2
gradient of a scalar function, 162
gravitational constant, 21
gravitational mass, 143
gravitational potential energy, 160
harmonic oscillator potential energy, 164
Hooke's law, 7
inclined plane problems, 256
incompressible liquid, 265
inertia tensor, 33
inertial frame, 90
inertial mass, 93, 143
instantaneous acceleration, 54
instantaneous axis of rotation, 219
instantaneous rest frame, 221
instantaneous velocity, 53
interconvertibility of mass and kinetic energy, 157
inertial and non-intertial frames/ pseudo-forces, 128
Kinematics, 53
kinetic energy, 150, 156, 157
kinetic friction, 167, 274
Kirchhoff's first law, 38
laws of motion for many-particle systems, 185
laws of reflection, 12
left-handed coordinate system, 49
length contraction, 147
linear density, 16
linear motion, 74
localized vector, 225
Lorentz force equation, 50
Lorentz transformations, 52, 146
Mach's principle, 138
magnetic dipoles, 103
magnetic field, 6, 28
magnetic forces, 91
magnetic moment, 143
mass defect, 184
Maxwell's equations, 144, 148
mole, 3
moment of inertia tensor, 228
momentum conservation, 99, 157
Mossabauer spectroscopy, 181
natural units, 4, 5
Newton's laws, 90, 92, 94
Newton's laws for extended bodies, 94
Newton's pail experiment, 131
Niels Bohr, 18
nuclear reactions, 184
nuclear strong force, 176
one-dimensional collision, 176
one-dimensional kinematics, 79
one-dimensional vectors, 23
order of approximation, 11
orthogonal coordinate system, 229
parallelogram law, 22
paraxial rays, 13
path length, 75
path independence, 159
Pauli, 178
photon, 180
pion, 176
Planck's constant, 20
plane polar co-ordinates, 257
point dipole, 46
polar vector, 48
position vector, 117
potential energy, 158, 160, 165
pressure gradient, 265, 267
pressure, 36
principal moments of inertia, 229
privileged frame, 133
projectile motion, 242
pseudo-forces, 92
pseudoscalars, 48
pseudovectors, 48
quadrupole moment, 231
quasi-static process, 155, 204
recoil velocity, 179

rectangular cartesian axes, 25
reflection of axes, 48
relativistic dynamics, 157
relativity of simultaneity, 147
rest energy, 156
rest mass, 148, 157
retardation, 58
retarding force, 151
right-handed coordinate system, 49
right-handed screw, 29
right-handed triad, 218
rigid body, 154, 215
rolling without slipping, 194, 221
rotated frame, 41
rotating frame, 218
rotational degrees of freedom, 227
rotational kinetic energy, 154
rotational law of motion, 231
Rutherford atom/model, 19
scalar equations, 151
scalar function, 161
scalar multiplication, 29
scalars, 22
second rank tensor, 44
self-adjusting force, 105, 112, 210, 247
shearing stress, 36
significant figures, 8
simple harmonic motion, 16, 83
simultaneous events, 122
size of an atom, 19
space-frame, 232
space-time coordinates, 146
special relativity, 101, 145
spherical polar coordinates, 253
spin, 234
spring force, 158
stability of equilibrium, 252
static equilibrium, 188
static frictional force, 139, 165, 247
strain, 36
susceptibility tensor, 33
system of particles, 154
temperature field, 28
tension, 91, 259
time dilation, 147
time-displacement invariance, 172
transformation of quantities, 51, 125, 117
translation and rotation, 119
transverse waves, 16
two-body decay, 178
uniform circular motion, 153
uniform translational motion, 153
unit vector(s), 126, 271
units, 1, 2, 6
unstable equilibrium, 254
variable acceleration, 65
variable mass system, 97, 99
vector addition rule, 24
vector algebra, 25
vector kinematic relations, 79
vector multiplication, 29
vector triple product, 228
velocity addition rule, 145, 147, 264
velocity of light, 145
viscous forces, 158
volume stress, 36
weak and strong form of the III law, 102, 187
weak interactions, 50
work–energy theorem, 150, 197